Microelectronics – Systems and Devices

Microelectronics – Systems and Devices

Owen Bishop

Newnes

OXFORD   AUCKLAND   BOSTON   JOHANNESBURG   MELBOURNE   NEW DELHI

Newnes
An imprint of Butterworth-Heinemann
Linacre House, Jordan Hill, Oxford OX2 8DP
225 Wildwood Avenue, Woburn, MA 01801-2041
A division of Reed Educacational and Professional Publishing Ltd

 A member of the Reed Elsevier plc group

First published 2000

British Library Cataloguing in Publication Data
A catalogue record for this book is available from the British Library

ISBN 0 7506 4723 X

Printed and bound in Great Britain

Contents

Preface

This book is written for a wide range of pre-degree courses in Microelectronic Systems. The contents have been carefully matched to current UK syllabuses at Level 3, but the topics covered, depth of coverage, and student activities have been designed so that the resulting book will be a student-focused text suitable for the majority of courses at pre-degree level around the world. The only prior knowledge assumed is basic maths and science.

The UK courses covered by this text are:

- Advanced GNVQ Units in Microelectronic Systems from Edexcel
- BTEC National units in Microprocessor systems, Micro-Electronic Systems, and Software Design Methods.

Essential theory is provided here but the book is strongly practical in its approach, encouraging students to assemble and test real microelectronic systems in the laboratory. The examination syllabuses do not specify which processors and which programming languages the student should cover. The suppliers' catalogues list several hundred microprocessors and microcontrollers and any one or more of these could be selected as a subject for study. There is likewise a variety of languages or versions of languages that may be used to program them. To keep the size of the book within reasonable bounds, the book looks at the Zilog Z80 as a typical microprocessor, and at the Atmel AT90S1200 as a typical microcontroller. Both of these are readily available from the major suppliers such as Farnell, RS Components and Maplin, as well as from several of the smaller firms. Other processors are mentioned where they show interesting differences from these two types. With regard to languages, the book concentrates on assembler (for the Atmel controller), BASIC and PBASIC (for the Stamp). Other languages are described, including C.

The descriptions of these processors and languages are intended to exemplify processors and languages in general. They are aimed at giving the student a wide view of the topic, but it is not expected that students will centre their studies on these particular processors or languages. In keeping with the syllabuses, the book leaves the student with an unrestricted choice of devices, prototyping systems and programming languages.

The book is a study guide, suitable for class use and also for self-instruction. The main text is backed up by boxed-off discussions and summaries, which the student may read or ignore, as appropriate.

There are frequent 'Test Your Knowledge' questions in the margins with answers given at the end of the book. Another feature of the book is the placing of short 'memos' in the margins. These are intended to remind the student of facts recently encountered but probably not yet learnt. They also provide definitions of terms, particularly of some of the useful jargon associated with microelectronics and computing.

Each chapter ends with a batch of examination-type questions, and in most instances with a selection of multiple choice questions. Answers to the multiple choice questions appear at the end of the book.

Owen Bishop

Part A – The Hardware

1 Systems in action

Summary

The essential features of a microelectronic system are described. These are illustrated by descriptions of typical systems: a cordless telephone, programmable logic controllers in industry, a personal computer, measuring instruments and data loggers, the control room of a power station, and distributed processing in flight control of aeroplanes.

Digital: a digital quantity (often a voltage) takes *either one* of two values. Contrast this with an *analogue* quantity (often a voltage), which takes *any* value within a given range.

Integrated circuit: one that is built up from a large number of components connected together on the same chip and contained in a single package. The term 'integrated circuit' is shortened to 'IC' in this book.

Program: a series of instructions telling the CPU what to do.

Microelectronic systems are *digital* systems, built from one or more *integrated circuits*. They contain a *central processing unit* (CPU) which is *programmed* to make the system perform its tasks. The CPU is either a microprocessor or a microcontroller or, in complex systems, there may be more than one. Microelectronic systems are widely used in equipment and installations such as washing machines, automatic teller machines, personal computers, production line packaging machines, printing presses and radio-telescopes.

The CPU has such complicated tasks to perform that is is made up of hundreds of thousands or even millions of transistors, as well as resistors and other components, all assembled on the same silicon chip. There are two main types of CPU:

- Microprocessors
- Microcontrollers.

These have much in common, although there are important differences between them. A micro*processor* is just what its name says. It is a data *processor*. It is designed to be able to *process* large quantities of complex data at high speed. It needs the support of other units, such as memory and input/output devices to make up a complete system. A micro*controller* usually operates more slowly and has less processing capability but it has the advantage of having the other units on the same chip. It is a 'computer on a chip' and, as such, is used on its own to take full *control* of a piece of equipment or installation.

Because CPUs are programmable, the behaviour of the system can be changed in many ways simply by revising the program. This gives microelectronic systems a big advantage over hardwired logical systems, such as are used in the less expensive home security systems, the older types of washing machines, and in remote control systems. Such systems may perform their intended task well but it is not easy to alter the system once it has been wired. Any major change of action usually involves rewiring, possibly changing some of the ICs and, frequently, making so many alterations that it is simpler to scrap the circuit and build a fresh one on a new circuit board. The flexibility of microelectronic systems is one of the main reasons that they are so widely used.

Test your knowledge 1.1

What are the four key features of a microelectronic system?

High and low

The two values that a digital quantity can take are often referred to as *high* and *low*. A high value represents logical high, which is equivalent to the binary digit '1'. A low value represents logical low and is equivalent to the binary digit '0'. Some systems operate with the values the other way round (negative logic) but this is very unusual.

In microelectronic circuits these values are represented by voltages. Low is often represented by 0 V, or a value fairly close to 0 V. High is often represented by +5 V, but other values may be used in other types of microelectronic system.

Note that, to a power engineer, *high* voltage means something more than mains voltage, for example 450 V, or even 132 kV. To a microelectronics engineer 'high' usually means a mere 5 V. It is slightly higher in certain kinds of system. However, high is just 3.3 V in the *low voltage* systems that are intended for portable battery powered equipment.

A survey

A simple system

Before we look in more detail at what what goes on inside a CPU, we will take a few examples of the way microelectronic systems are used in everyday life and at work. The cordless telephone is a typical example (see over). As in any other microelectronic system, the circuit centres on a CPU. In a cordless telephone this is a microcontroller, complete with memory.

It may be wondered why a cordless telephone needs a CPU, yet an old-fashioned corded telephone can operate without. One of the reasons is that the corded telephone is wired directly to the public network, but the cordless telephone has to make radio contact with its base before a call can be received or made.

The handset of a cordless telephone (Fig. 1.1) consists essentially of a radio transceiver that is under the control of a CPU. The radio has limited range and communicates with the base unit, which may be up to 200 m away, normally in the same building. The base unit

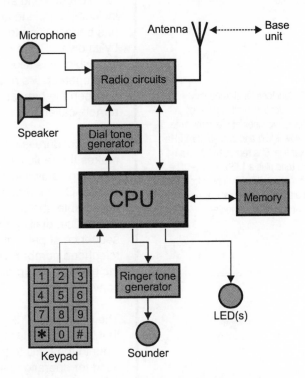

Figure 1.1 *A cordless telephone handset is under the control of its central processing unit (CPU). This may be a microprocessor or (more often) a specialised microcontroller.*

The exchange of signals (such as identity codes, 'ready to receive' , 'ready to transmit') between the handset and its base station is known as *handshaking*. A similar handshaking dialogue occurs when a microcomputer communicates with a printer.

DTMF: dual tone multi-frequency dialling codes a dialled number by producing two tones at the same time, one below 1 kHz and one above 1 kHz. For example '5' is represented by a 770 Hz tone plus a 1336 Hz tone. The system codes all the numbers from 0 to 9 and also 'star' and 'hash' by using 12 different pairs of frequencies.

Cordless telephones

Fig. 1.1 shows that the radio circuits are under the direct control of the CPU. The double-headed arrow between the CPU and radio circuits indicates that signals may pass in both directions (though not both ways at exactly the same time).

The radio circuits of the base station and the handset are permanently switched on, waiting for a call. When the base station receives an incoming call from the public network, it sends out a digital signal by radio. This signal includes a code that identifies the base station. The handset receives this signal. However, it is also able to receive signals from any other base stations within range. The signal is sent to the CPU, which then checks to find out if this is the code of its own base station. If it is not, it ignores it and nothing further happens. If it is recognised, the CPU makes the radio circuit send an acknowledging signal. The signal includes a code to identify the handset. The base station has been waiting to receive this signal, which is checked by its own CPU to make sure that it comes from a handset with which it is allowed to communicate. Then the CPUs both open up the radio channels for two-way conversation between the handset and the base, and by land line to the remote caller.

The procedure is similar for an outgoing call. The CPU makes the radio circuit transmit a series of code groups, including its own identity code and the number to be dialled. On confirming that the identity of the handset is acceptable, the CPU of the base station dials the number. In practice, dialling the number means generating a sequence of pairs of DTMF tones that code the required telephone number. When the number answers, it replies to the handset and radio channels are opened for two-way conversation.

The lower part of the diagram shows the input and output that links the CPU to the user. The keypad is used to send input to the CPU to tell it what number to call. It is also used for operations such as storing frequently needed numbers in memory. The CPU has output to one or more signal LEDs that indicate when a call is in progress. It has output to a separate IC which includes an oscillator to generate the ringing tone.

communicates with the public telephone system through the subscriber's ordinary telephone line. The circuit of the base unit is similar to that of the handset in many ways.

The operating system is stored permanently in a part of memory when the telephone is manufactured. There is also a section of memory to hold useful data, such as the number currently being dialled and a list of frequently used telephone numbers. This data is changed from time to time by the user.

Operating system: a program which tells the CPU how to do all the basic tasks of running the system.

A necessary part of any microelectronic system is the squarewave generator known as the *system clock*. This provides the regular series of pulses that drive the CPU. It is not shown in Fig. 1.1 because it is usually included on the same chip as the CPU. The timing of the clock usually depends on a quartz crystal, just as in a digital watch. There is no room for the crystal itself on the CPU chip, so this is connected across a pair of terminal pins of the IC. The frequency of the crystal may be several hundred kilohertz or a few megahertz.

One of the essential outputs of a telephone is the ringing tone. It would be possible for the CPU to be programmed to generate this tone itself, but generating the tone would occupy the CPU at times when it could more usefully be doing something else. It is common in microelectronic systems to employ special-purpose ICs like this where there are simple repetitive tasks to be done. The telephone has another special IC to generate the DTMF dialling signal for transmission to the base station.

Test your knowledge 1.2

Why is it preferable to have a DTMF generator in the telephone handset?

Test your knowledge 1.3

What is the name given to the square wave generator that drives the CPU?

Cellphones have circuits similar to cordless phones, the main difference being that the cellphone communicates directly with the public system through a base station up to several kilometres distant. There is usually an LCD message screen to display numbers dialled and other useful information.

Controllers in industry

Microelectronic systems are widely used in industry. This section describes an example of microelectronic control of a chemical process (Fig 1.2) by a *programmable logic controller*, or PLC. The CPU (with system clock), its memory, keypad and display, are part of a single unit (Fig 1.3). As in the telephone, the heart of the system is a CPU. This has access to memory for storage of the program and working data. In some systems the whole memory or part of it is included on the CPU chip. There is often a keypad by which the operator runs the system, and there is a message panel on which the CPU displays information about the current state of the system.

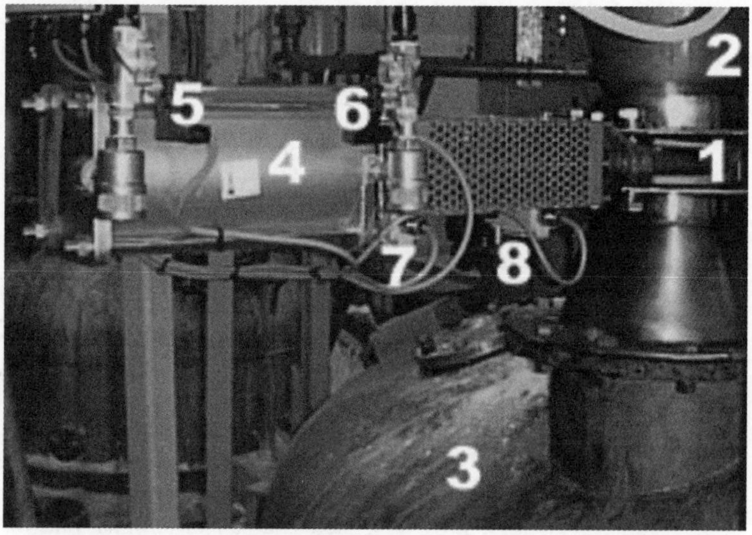

Test your knowledge 1.4

What device is at the centre of the system of a programmable logic controller?

Figure 1.2 *In the manufacture of chipboard, the bonding resin is made by heating a mixture of urea and formaldehyde. A slider valve (1) controls the flow of urea from a hopper (2) to the processing kettle (3). The valve is opened or closed by a shutter that is moved by a piston enclosed in a cylinder (4). The piston is moved by admitting compressed air into the cylinder on one side of the piston or the other side. The flow of air is controlled by two solenoid-operated air valves (5 and 6), which are switched on or off by the microcontroller. Proximity sensors detect when the valve is fully open (7) or fully closed (8) and supply this information to the microcontroller.(By courtesy of Kronospan Ltd., Chirk)*

Actuator: A device in a control system which performs an action. Examples are motors and solenoids or devices driven by motors or solenoids, such as valves.

In the cordless telephone previously described, currents are small and can usually be fed directly to the inputs of the CPU. Similarly, the outputs of the CPU can provide sufficient current at the correct voltage to drive logic circuits, including those driving display circuits and tone generators. This is rarely the case in industrial plant. Motors often operate on a 24 V DC supply or even run on alternating current at mains voltage. Similarly, signals from sensors may be at voltages higher than those acceptable by the CPU, and may sometimes be AC signals.

Interface: A circuit used for passing information between two other circuits. Examples: a modem passing digital data between a computer and the telephone network, an opto-isolator passing simple 'go-stop' commands between a CPU and sensors or actuators.

Industrial sites are well known for generating strong electromagnetic interference, so the input signals from sensors may carry high voltage spikes. EMI may also be picked up by the output circuits and could get back to the processor. For this reason, interface circuits (see Fig. 5.1) are needed, both on the input and output sides to provide a low-voltage, low-current, electrically 'quiet' environment in which the CPU can operate reliably.

A Seimens 95U PLC can be seen in Fig. 1.4. The PLC is wired to a number of input and output interfaces which are mounted on the rack in the cabinet. Cables run from these to the sensors and actuators on the plant. A few others run to control switches and indicator lamps on the door of the cabinet. The door is normally closed when the plant is operating, so acting as a control panel.

The program of a PLC runs continuously in a loop for as long as it is switched on. The first stages of the program read the state of each sensor and store the results in a special area of memory. Then the program examines the input data and decides what action is to be taken. As an example, take the valve mechanism of Fig. 1.2. If the proximity sensor (7) shows that a shutter has reached the far end of its travel, the valve (5) admitting compressed air to the nearer side of the cylinder must be closed. A message indicating 'close valve' is stored in the output area of memory of the PLC. When all the logical decisions have been taken and the future output state is stored in memory, the program reaches its third and final stage. It sends the stored output data to the actuators. The actuators are switched on or off in response to the latest state of the system. The program repeats immediately, so it is

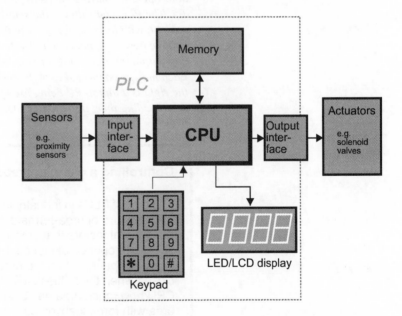

Figure 1.3 *Like almost every other control system, a PLC is centred on a CPU. The dotted line indicates that the CPU, memory, keypad and display are normally installed as a single general-purpose unit. Interfaces to sensors and actuators may be separately installed, and there may be several hundred in the system. Smaller systems may use PLCs with a dozen or so built-in interfaces.*

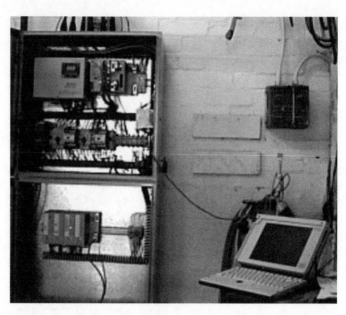

Figure 1.4 *A PLC system is housed in a cabinet, shown here with the door open. It controls the resin production plant illustrated in Fig. 1.2. The PLC controller is the small box mounted at the top left of the cabinet, with its control keys situated below its LED display. The low-voltage power supply is mounted to the right of the PLC. The input and output interface units are mounted on the rack below the PLC. Each may be connected to up to eight sensor or actuator devices. On the right is a laptop PC being used for writing programs and downloading them into the PLC. (By courtesy of Kronospan Ltd., Chirk)*

Controlling a chemical reaction

The use of PLCs in industry is illustrated by a stage in the manufacture of urea-formaldehyde resin (Fig. 1.2). There are several factors that determine whether the shutter should be opened or closed. For example, the shutter must be opened when the process begins, and must be closed when the kettle is full. Weighing sensors tell the CPU how much urea has been added to the kettle. Mixing urea with formaldehyde causes heat to be generated so a thermal sensor provides essential input to the CPU. The rate of addition of urea must be carefully controlled so that it does not overheat. The CPU controls another actuator which is a water valve which admits cold water to pipes surrounding the kettle. The program continually checks temperature and adjusts the rate of addition of urea and the rate of flow of cooling water accordingly.

continually reading input from the sensors, taking decisions and sending the appropriate output to the actuators. A typical program has a few hundred or thousand steps and take only a few tens of miliseconds to run, so the system responds reasonably quickly to changes in the state of the inputs.

The example of the valve demonstrates that it is not enough for the CPU to instruct actuators to move the shutter. There must also be sensors to check that the shutter is actually open or shut. This is to allow for the fact that it may not have had time to move to the required position. Or maybe it has jammed.

A system of the kind shown in Fig. 1.3 is common in industrial plant, whether it is a simple machine for filling cartridges with toner powder, a vast printing press, or a chemical plant producing insecticides. The main difference from one system to another is in the types and numbers of sensors and actuators attached to the system. The other main difference is the program that directs it. The program for the PLC is written by the operator, using special software running on a microcomputer. The PLC in Fig. 1.4 was in the process of being programmed by the laptop PC on the right. The program is tested on the microcomputer and, when it is free from bugs, downloaded into the memory of the PLC. Once the program is running correctly, the PC is disconnnected from the system and the PLC runs independently. The program is not normally altered except when there is to be a change in the operating procedure of the plant.

Bug: a programmer's term for an error in a program.

PCs and similar computers

There is much to be said about computer systems in later chapters, so for the present we will simply state the main ways in which they differ from the typical microcontroller systems described above. In essence, all computers have the same main features and we may take the typical personal computer (PC), as our example (Fig. 1.5).

The basic features are the same as in any microelectronic system: CPU, memory, input and output. Because the PC is intended to perform a wide range of often complex operations at high speed, a microprocessor is chosen as its CPU. Usually the system clock is a separate unit, as shown in Fig. 1.5. In contrast to systems such as the cordless telephone and PLCs, the PC has a full-sized keyboard, with over a hundred keys. It normally has a colour monitor.

The PC has several other input and output units either built into it or connected by special sockets. These include disk drives of various kinds, a mouse, and a printer. There may also be other devices such as

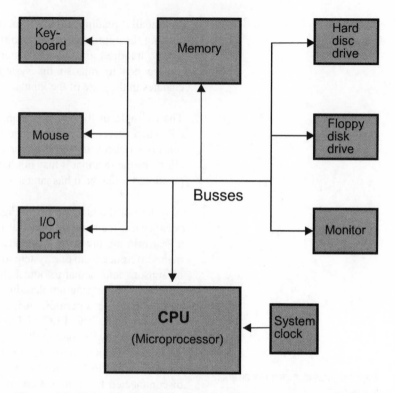

Figure 1.5 *This simplified diagram of a PC shows that it has much in common with the typical control systems of Figs. 1.1 and 1.3.*

a pair of loudspeakers, a joystick, a scanner and a digital camera.

One of the distinctive features of a PC and other computer-like systems is the *bus*. To assist the rapid transfer of data between the CPU and the other parts of the system the units are linked by a set of parallel conductors, shown for simplicity as a single conductor in Fig. 1.5. In practice, the bus consists of three separate busses, each with its own task, as will be explained in Chapter 2.

A PC is programmed from various sources. First of all, it has a block of permanent memory in which the operating system is stored. It has numerous programs stored on its disk drives, and the user can purchase other programs on compact discs, or download them from the Web. These programs are temporarily transferred to the computer's memory when they are to be run. Programs include word processors, spreadsheets, accounts programs, games, educational and training programs, and information programs such as dictionaries, encyclopaedias, telephone directories, catalogues and atlases. A wide range of specialised programs is obtainable for use by travel agents, theatre booking agents, medical centres, libraries and other medium sized organsations. Major businesses and organisations such as banks and oil companies employ software writers to produce programs intended for their operations (Fig. 1.6).

Figure 1.6 *Using a mouse to control a power station. The state-of-the-art control room at Ironbridge Power Station, Shropshire, uses computer monitors to display data readings taken at dozens of points in the steam-producing plant, in the turbines and in the electricity generators. The operator controls the power station by calling up a virtual control panel on the monitor. On this, the usual control switches, variable resistors, indicator lamps and meters are displayed in diagrammatic form. There is no keyboard on the computer. Instead, the operator uses a mouse to operate the controls, clicking on 'buttons' or dragging 'sliders' in just the same way as when playing a computer game. The station also has a basic conventional control panel with real switches for back-up in case of computer failure. (By courtesy of Eastern Generation Ltd.)*

Measuring instruments

Except in the cheapest models, the circuit of a digital multimeter (Fig. 1.7) centres on a microcontroller. This makes it possible for the multimeter to perform a range of functions quite beyond the scope of the conventional analogue multimeter, based on a moving-coil microammeter. The user of the digital meter simply selects what quantity is to be measured, applies the two probes to two test points in the circuit, and the reading automatically appears on the display. It is automatically updated several times per second. The display usually consists of 4 digits, with a movable decimal point and a polarity indicator (– for negative values). The quantities that can be measured include voltage, current, resistance, capacitance and frequency. With an ordinary multimeter the user has to select the range, but range selection is automatic with a meter based on a microcontroller.

Figure 1.7 *A multimeter is no longer a switched network of resistors and capacitors connected to a sensitive moving-coil microammeter. It still has the resistor and capacitor network, but now most of the switching is done by CMOS gates managed by a microcontroller.*

Given a constant current source, a timing circuit and a voltage-measuring circuit, it is possible to measure all the other quantities. Resistance can be measured by finding the voltage drop across the resistor when a given current flows through it. Capacitance can be measured by finding how long it takes a constant current to charge the capacitor to a given voltage. Frequency can be measured by timing the changes in voltage. These tests are easily automated for different ranges.

One of the few disadvantages of the numeric display of a digital meter is that, with a varying quantity, the rapidly changing figures do not help the user to visualise the way in which the value is changing. The needle of the conventional electromagnetic meter is much better for this purpose. The multimeter makes up for this with a bargraph display, which can be seen running along the lower margin of the display in Fig. 1.7.

The meter can also process the measurements. If the meter is run for a few minutes or more, the user can view the values as they change, or the microcontroller can be programmed to pick out the maximum value and the minimum value, and to calculate and display the difference between maximum and minimum, and the average value. It can also measure the voltage produced by a thermocouple and calculate the equivalent temperature in Celsius or Fahrenheit.

The next grade of microelectronic instruments above the multimeter is the *data logger*. In practice, these instruments perform two related but distinctive tasks:

- *Data acquisition* – receiving data (voltages, counts) from sensors.
- *Data logging* – recording data and processing it.

The Datataker (from Data Electronics), upon which this descripton is based, acquires measurements from a number of sensors connected to its array of input terminals. The measurements may be displayed on the Datataker's own screen or on the monitor of an attached computer. Subsequently the data can be stored (logged) in its own memory or on removable memory cards. The data can then be processed. For example, the device can calculate maxima, minima and other functions. It can also convert voltage, for example, into temperature in Celsius or Fahrenheit. The Datataker is able to perform its calculations at a higher level than the multimeter. For example, it has selectable routines for different types of thermocouple instead of being restricted to the one type supplied with the instrument. The scope of processing is increased by including statistical operations such as calculating standard deviations and plotting histograms. There are many other refinements in data presentation. For instance, each measurement is 'time and date stamped' with the time and date at which it was taken.

Another major bonus of the data logger is that it is programmable. It can be set to take periodic readings from a number of different sensors and store the results for display later. Or it may be set to produce an alarm output when values fall in a specified range. The data logger is programmed by using word-processing software running on a PC. It has its own specialised programming language. The finished program is downloaded from the PC into the Datataker. This can then run the program on its own, when it is no longer attached to the computer.

Distributed processing

In a conventional microelectronic system the CPU has direct control over all the input and output devices in the system. Figs. 1.1, 1.3 and 1.5 show examples of this. Now that a wide range of microcontrollers is available cheaply, a new approach to control has become more widespread. Each sensor or actuator in the system has its own microcontroller as an integral part of it. The microcontroller is programmed specifically to manage the action of the sensor or actuator. For example, consider an electric motor geared to an aileron in the wing of an aeroplane. When the aileron is to be moved, a command is sent from the central computer in the pilot's cockpit to the controller in the wing. The controller is then responsible for moving the aileron to its new angle. Normally it will accelerate it as fast as mechanical stresses allow until it reaches the maximum allowable rate of turning. This controlled acceleration involves complicated calculations by the microcontroller, based on previously determined parameters. There must be feedback of the actual position of the aileron to allow for mechanical effects such as wind resistance.

The controller has been told the angle at which the aileron is to finish so, *before* it reaches that position, the controller begins to decelerate it at the maximum allowable rate so that it finally comes to rest at exactly the required angle. While the action is in progress and when it is completed the processor reports back to the main computer. The main computer may also interrogate it at any stage to find out what angle the aileron has reached.

This is a reasonably complicated operation in which factors such as wind resistance must be taken into account. It is simpler for the task to be undertaken by an independent processor situated at the motor, than it is for all the ailerons and other control surfaces to be controlled from the central computer. In addition, this approach requires less cabling and is less subject to electromagnetic interference.

Distributed processing is part of the new 'fly by wire' principle adopted by Lucas Aerospace, as used in the A320 'Airbus' aeroplane.

Summing up

Microelectronic systems may be divided into::

- Control systems – examples: resin production, flight controls, automotive applications.

- Instrumentation systems – examples: testmeters, data acquisition and data logging.

- Communications systems – examples: cordless telephone, facsimile machine.

- Commercial systems – example: personal computer, automatic teller machine, EFTPOS station, stock logger.

Activity 1.1 Microelectronic systems

Find out as much as you can about a machine or equipment that is based on a microelectronic system.

Examples include:

Washing machine
Fax machine
Telephone answering machine
Dot-matrix printer
Stock logger
'Smart' room heater
Compact disc player
Garden or greenhouse reticulation (watering) system
Car park entry and exit control
EFTPOS machine (Electronic fund transfer at point of sale)
ATM (Automatic teller machine, also known by other names such as cash dispensers)
Global positioning system device
Radar systems (including radar speed traps)
Traffic control systems
Car engine control
Robotics

Sources of information are:

Books in the public library and your departmental library
Back issues of technical periodicals

Manufacturers' advertising matter and brochures
Manufacturers' and other sites on the Internet
Arranged visits to local factories and businesses

Problems on systems in action

1 List the stages in making a call from a cordless phone, referring to the parts of the system that are pictured in Fig. 1.1. Cover the action from the time the handset is switched on until the first words are spoken.

2 Outline the structure and action of a programmable logic controller.

3 Explain why special interfaces are needed between a PLC and the attached sensors and actuators.

4 Describe the sequence of actions as a PLC runs its program.

5 In what ways does a digital multimeter based on a microcontroller differ from an analogue multimeter with a moving-coil microammeter? What are the advantages of the digital multimeter?

6 List the devices (peripherals) that may be attached to a PC and explain briefly what they do.

7 Describe the features and action of a data logger.

8 What is meant by *distributed processing*?

9 Write an essay under the title 'Microelectronic systems in everyday life'.

Multiple choice questions

1 Lamps, motors and solenoids are examples of:

    A sensors.
    B interfaces.
    C actuators.
    D outputs.

2 A CPU is:

    A a microelectronic system.
    B the heart of a microelectronic system.
    C unit which stores a program.
    D a computer on a chip.

3 When a PLC is running its program it is directly connected to:

    A sensors.
    B actuators.
    C a PC.
    D interfaces.

4 A microcontroller:

    A has its CPU and memory on the same chip.
    B has only a CPU on its chip.
    C controls a PC.
    D is designed to process data at high speed.

2

The CPU

Summary

The controlling centre of a microelectronic system is its central processing unit or CPU. A typical microcontroller, the Atmel AT90S1200, and a typical microprocessor, the Zilog Z80, are examined in detail, and compared with other popular units of the same type. The architecture of microcontroller systems is compared with that of microprocessor systems. The bus of microprocessor systems is seen to comprise a data bus, an address bus and a control bus. The functions of each are described. In connection with the data bus, the purpose of three-state outputs is described. Discussion of the address bus centres on address decoding. The functions of the signals on the control bus of the Z80 are explained. The system clock co-ordinates the activities of the CPU and the other units in the system.

As we have seen in Chapter 1, a microelectronic system has a number or fairly standard parts:

- the CPU.
- the system clock.
- conductors for carrying signals between the CPU and the other parts of the system.
- memory, of various kinds.
- an assortment of input and output circuits and devices, mainly depending upon the application.

The controlling centre of a microelectronic system is the CPU. Its function is to read data from certain parts of the system, to act on it (process it) and to output the results. As explained in Chapter 1, microcontrollers differ in several ways from microprocessors, so we consider them separately. There are several hundreds of different microcontrollers and microprocessors. In this chapter we consider a few typical examples.

Microcontrollers

As our first example of a CPU, Fig. 2.1 shows a commonly used microcontroller, the Atmel AT90S1200, which we shall refer to as the '1200' for short. Remember that this is a complete system on a single chip so its use of terminal pins (its *pinout*) is very different from that of the microprocessors that we describe later. This simplicity is what makes systems based on microcontrollers so much easier to design and build. The terminal pins fall into four main groups:

- power lines – the positive (V_{CC}) and 0 V (GND, short for ground) lines connect to these pins. The voltage between them should be in the range 2.7 V to 6 V.
- crystal – the system clock circuit is inside, except for the crystal, which must be connected across these two terminals, XTL1 and XTL2. The maximum crystal frequency is 16 MHz.
- I/O ports – there are two of these, port B and port D (larger 40-pin members of this series also have ports A and C). Port B has 8 pins, PB0 to PB7, while port D has only 7 pins, PD0 to PD6. The 'bits' of each port (the individual pins) can be programmed as inputs or as outputs.
- reset – this line is held high when the CPU is running. A low level (0 V) on this line resets the system.

I/O: input/output.

A line drawn over any symbol is used to indicate that it is *active low*. For example, the RESET input is normally held high, but resets the system when it is made low.

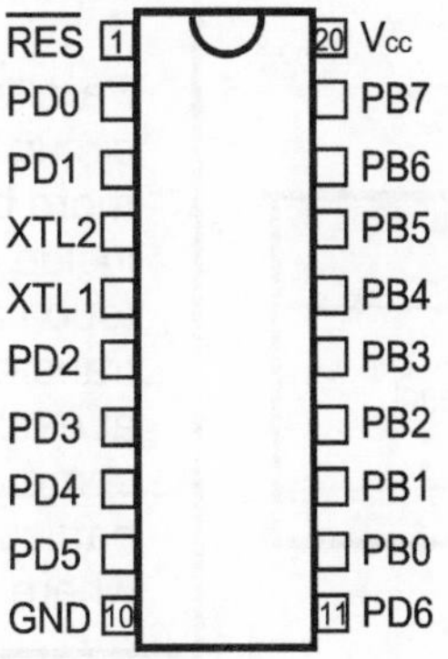

Figure 2.1 *The pinout of the 1200 illustrates the way we number the pins of ICs when there is a row of them down each side of the IC. As viewed from above (pins pointing down to the circuit board) pins are numbered from 1* down *the left side, continuing* up *the right side. A notch indicates the end of the IC where pin 1 is located. Some makers mark this end with a dot or a stripe instead.*

Bit: short for binary digit. A bit takes one of two values: 0 (equivalent to low or OFF) and 1 (equivalent to high or ON).

Nybble: a binary value consisting of 4 bits.

Byte: a binary value consisting of 8 bits. Its value can range from 0 decimal (0000 0000) to 255 decimal (1111 1111).

Word: a double-byte of 16 bits. Its value can range from 0 decimal to 65565 decimal.

Numbering digits

When a number consists of more than one digit, they are numbered from right to left, starting with number 0. For example, in the number 5239, D0 is '9', D1 is '3', D2 is '2' and D3 is '5'. D0 is always the least significant digit (LSD). In a 4-digit number, D3 is the most significant digit (MSD).

In the '1220' IC (Fig. 2.1) the eight terminals of port B are numbered from PB0 (LSD) to PB7 (MSD), using the same rules.

Parallel and serial data

Digital data are transferred from one place to another either in parallel or in series.

Serial transfer (Fig 2.2) requires only one line, along which the voltage levels are sent one after the other.

In parallel transfer (Fig. 2.3), a binary number is represented by high and low voltages placed on a set of conductor lines at the same time. Transfer requires a separate line for each bit.

Parallel transfer is quicker but requires more tracks on the circuit board, more terminals on the ICs and may require more buffers at each stage to relay the signals. It is used inside computers and other data-processing equipment (such as modems) to provide speed. In contrast, serial transfer needs only one track, one terminal and one buffer at each stage, but it is much slower. It is often used for communicating between microelectronic systems, particularly by the telephone system or by radio links, where only one channel is available.

What is the advantage of serial transfer of data?

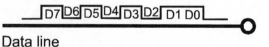

Data line

Figure 2.2 *Serial transfer of a byte of data uses only one data line but takes about 8 times as long as parallel transfer.*

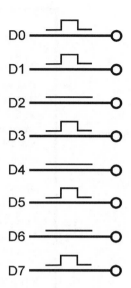

Figure 2.3 *Parallel transfer of 1 byte (8 bits) of data requires 8 data lines, D0 to D7.*

In order to keep the size (or more exactly, the number of pins) of the IC reasonably small, manufacturers often allocate two or more functions to the same pin. In the instance of the '1200' the I/O pins PB5, PB6 and PB7, together with the RESET pin, can also be used as a serial port for downloading a program from a computer (such as a PC) to the memory inside the '1200'. In serial transfer of data, high or low pulses are fed into or out of the pin one after another (Fig. 2.2).

It can be seen that the pins of the '1200' provide all the essential access to the system: a power supply, a timing crystal (too big to go on the chip), a way of quickly resetting the system, and connections for sensors and actuators for inputting or outputting data.

Data can be fed to the '1200' in parallel (Fig. 2.3). That is, we simultaneously send a '1' or '0' along its 8 Port B lines and so load it with 8 bits (a byte) of data in a single operation. We can do the same thing with Port D, except that this has only 7 bits. Similarly, we can read (in parallel) a byte from Port B or a 7-bit group from Port D.

As might be expected from their name, *microcontrollers* are mainly used in control applications. A port may receive a byte of data from a temperature sensor, the value of the byte representing the temperature. In the other direction, the port may output a byte of data to control the speed of a motor. However, the input from a sensor may not be a byte but a single bit.

These types of memory are described in Chapter 4.

RISC: reduced instruction set computer.

Watchdog timer: a device for checking that the CPU is operating correctly.

Other controllers – the PIC family

The Microchip PIC family of microcontrollers has many and varied members. Some of the smallest are in a standard 8-pin IC package (Fig. 2.4). The 12CE518 runs on 2.5 V to 5 V, with a frequency up to 4 MHz. It has a program memory (PROM or EPROM) of 512 bytes. It also has 25 bytes of RAM and 16 bytes of EEPROM, which can retain stored data for as long as 40 years.

Programs are loaded serially and there are six I/O pins. Members of this family are RISC processors. The 12CE518 has only 33 instructions in its set, so learning to program it does not take long. All of its instructions except branches take 1 μs to execute. Branches take 2 μs.

It has a built-in timer that can function as a real time clock. It also has a watchdog timer that has its own oscillator to ensure reliability.

Another RISC (35 instructions) microcontroller is the 16F872, which is typical of the more advanced members of the PIC family. It is in a 28-pin package, which allows it to have three I/O ports that are 6, 8 and 8 bits wide. It runs at 20 MHz. It has 2K × 14 words of FLASH reprogrammable program memory, 128 × 8 bytes of RAM and 64 × 8 bytes of EEPROM. On the same chip there are three timers, a watchdog timer with its own oscillator, a 10-bit analogue to digital converter, a synchronous serial port. It can be programmed to capture a 16-bit value at regular intervals, which gives it application in data acquisition. The captured data can be compared with a value in another register and produce an output signal if the two are equal. There is also a pulse width modulator to generate pulses of a set length. In total, this is a very versatile controller with many applications in control systems and measurement systems.

There is a wide range of development equipment and software to help the PIC programmer. These include assembler programs and software for programming in the C or BASIC languages. The Stamp 2 is a complete microcontroller system based on the PIC16C57. It includes a compiler for PBASIC, which is a version of BASIC devised for making best use of the features of the 16C57. Some programs for the Stamp 2 are provided later in this book.

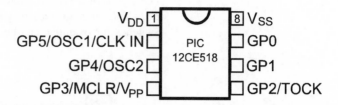

Figure 2.4 *The smaller PIC microcontrollers have only 8 pins. Some pins share several functions.*

Example:

In an industrial plant, either a valve is closed or it is not closed. The sensor is a microswitch on the valve, which closes when the valve is closed. The signal is sent from the microswitch to one of the port pins of the CPU. It is a high voltage (logic 1) when the valve is closed and is a low voltage (logic 0) when the valve is not closed. The signal is a single bit.

Microswitch: A light-duty switch which 'clicks' on or off when the pressure applied to its button or lever changes only slightly.

Example:

Either the temperature of a vessel has reached the required level or it has not. We can represent these conditions by input from a thermal sensor at just one pin of Port B. The input is high when the temperature has reached the required level and is low when it has not. Again, the signal is a single bit.

The same applies to many actuators. If a motor is either running or not running, only a single bit is required to switch the motor on (1) or off (0). For this reason, the individual bits of each port may be configured as inputs or outputs, and the output and input values set individually. On the '1200' we could have up to eight different sensors, or eight different actuators (or a mixture of both) connected to Port B and seven more connected to Port D.

When it is important to keep the size of the IC (and therefore the number of pins) to a minimum, it is possible to transfer data serially into or out of the microcontroller. Then we need only a single input/output pin. An example of this is seen in the '1200' microcontroller. In programming mode, the pin PB5 (see Fig. 2.1) is used for serial input of programs. Pin PB6 is used for serial output, when the program is read back into the programming PC to verify it. PB7 is used as a clock input for the serial data transfers. The main disadvantages of serial data transfer are that it takes longer than parallel transfer does and that the programming is more complicated.

The layout of a typical microcontroller system is represented in Fig. 2.5. The input devices may be directly connected to the I/O pins. Switches for example, can be wired directly to the IC as in Fig. 2.6.

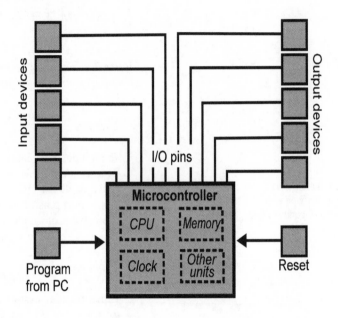

Figure 2.5 *A microcontroller has the CPU and most other units of the system on a single chip. It communicates with the external world through the I/O pins of its one or several ports. The input and output lines radiate from the microcontroller.*

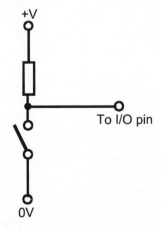

Figure 2.6 *A simple one-bit input to a microcontroller. The resistor holds the pin high (logic 1), except when the switch is closed, when the input is low (logic 0). A suitable value is 10kΩ. Note that +V is the processor supply voltage.*

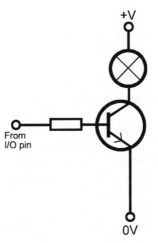

Figure 2.7 *Simple one-bit control of a filament lamp by a microcontroller. The resistor limits the current drawn from the output pin to a safe level. +V may be higher than the processor supply voltage, if required.*

Other input devices, such as photocells or Hall effect magnetic detectors, need a simple circuit to interface them to the microcontroller. Similarly, the pins of typical microcontrollers are able to source a few tens of milliamps, so indicator devices such as LEDs can be driven directly from the pins. Devices that will accept logic-level inputs can also be driven directly. Other output devices, such as solenoids and motors, require heavier currents and so need special interface circuits, such as a switching transistor (Fig. 2.7).

The '1200' is the simplest member of the AVR family of microcontrollers. Others have larger on-board memory and additional features such as a serial port, a pulse width modulator, capture/compare circuitry and an extra timer.

Microprocessors

At one time, the Z80 was used as the CPU in several of the popular hobby computers. It still has many applications in microelectronic systems, particularly those that require more computing power than a microcontroller is able to provide. However, as the CPU of hobby computers and other desktop computers, it has been superseded by much more powerful processors, such as the Intel Pentium and the AMD Athlon.

The Z80 is contained in a 40-pin package. Its pins fall into seven main groups:

- power lines – the Z80, like most microprocessors and other devices in the system, runs on a 5 V supply.
- data inputs/outputs – there are 8 of these, allowing a byte of data to be read or written by the IC.
- address outputs – there are 16 of these, providing 64K (see box) locations in address space. More advanced versions of the Z80

such as the eZ80 have 24 address outputs.
*   system clock input.
*   reset input.
*   control inputs – there are six of these. The inputs control the rate of processing or allow processing to be interrupted or slowed.
*   control outputs – there are seven of these, used by the CPU to communicate with the rest of the system.

The internal structure and activities of the microprocessor are described in Chapter 4. Here we look at the ways in which the microprocessor communicates with the other parts of the microelectronic system.

Architecture of the system

A microcontroller system usually has I/O connections radiating to the input and output devices (Fig. 2.5). By contrast, a microprocessor system has a single, central connector with branches going to the microprocessor and all other units in the system. This arrangement of connectors is known as a *bus*. The bus was first referred to in Fig. 1.5 but, in the more detailed drawing of Fig. 2.8, it can be seen that there are actually three busses running side by side. Each bus is connected to all the units of the system. The three busses are concerned respectively with data, addresses, and control (match this statement with the list of pin types in the previous section). Note that the data bus carries signals from the CPU to other parts of the system and also carries signals to the CPU from other parts. There is two-way communication. The address bus and individual lines of the control bus communicate in only one direction.

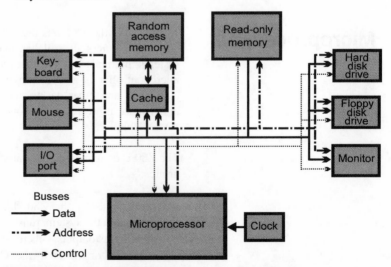

Figure 2.8 *A typical microprocessor system has all its parts connected to the data, address and control busses.*

Continual upgrading

The reader should be aware that the microelectronics industry is forever active in producing new devices and improving the performance of the old ones. It is impossible for a book to be really up-to-date. Specifications can and probably will change between the date this book is written and the date that the reader gets it, even though the dates are less than a year apart.

The best way for the reader to keep up-to-date is to study the various electronic magazines and to browse the electronic sites on the Internet.

Arrangements of pins

In the Z80, the terminal pins are arranged in two rows along either side of the ICs, as in Fig. 2.1, except that there are 40 pins (more in some versions). This is known as the *dual-in-line* (DIL) layout. CPUs that are more powerful have many more pins. This is mainly because of extra pins in the data I/O and address output groups. The extra data pins (bringing the total to 16 or 32) connect to additional circuitry on the chip. They allow the CPU to deal with much larger numbers and longer codes. The extra address output pins allow the CPU to address a much larger memory (see box, p. 30). The DIL layout is not suitable for more than about 40 pins. ICs with larger numbers of pins are usually square, with the pins arranged along all four sides in a single or double row. The most powerful CPUs, such as the Pentium and Athlon, have hundreds of pins and special layouts are used.

The data bus and three-state outputs

If we look even more closely at a bus, we see that each bus consists of many tracks running side by side. Fig. 2.9 shows a data bus that is eight bits (1 byte) wide. This is typical of most microprocessors. Voltages representing logic low (0) or high (1) are placed on this bus, by the CPU or other devices in the system. They represent a number or an operational code. This is how the CPU reads data from units such as the memory, or sends data to units such as an output port.

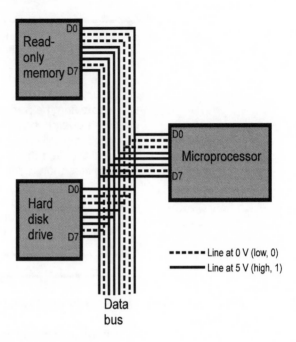

Figure 2.9 *A bus consists of a number of conductors running side by side on the circuit board. Here we illustrate a data bus that is 8 bits wide. The figure shows the voltage levels on the data bus when the binary value 1011 1001 is present.*

One essential point about a data bus is that only one unit can be allowed to place data on it at any one time. It is the same situation as a discussion group. If one person is talking, the others must keep quiet so that person can be heard by all. However, in all the units in Fig. 2.9 the data output pins are *permanently soldered* to the bus. It is inevitable, given the way in which microelectronic systems are constructed. The difficulty is that, even if the outputs of the 'quiet' units are at zero volts, they will pull the bus voltage levels down. The signal from the 'talking' unit will be 'swamped' by the 0 V levels from the 'quiet' units. The solution to this problem is for all units connected to the data bus to have *three-state* outputs. The three states of such an output are:

- logic 0 – the pin is at logic low voltage.
- logic 1 – the pin is at logic high voltage.
- the pin is in the *high impedance* state.

In the first two states the pin is electrically connected to the bus so that the output voltage (high or low) of the internal logic circuit appears on the bus. In the third state, a very high resistance is switched in, between the pin and the internal circuit. The resistance (or impedance) between the pin and the circuit is very high. In effect, the pin becomes

---

Bus width

A typical microprocessor such as the Z80 has eight lines in its data bus. We say the bus is eight bits *wide,* and we number the bits from D0 (the LSB) to D7 (the MSB). However, a bus with only 8 bits can transmit only the integer values from 0 to 255. It takes more bits to represent larger numbers or to represent quantities more precisely. The main way to allow for larger numbers and greater precision is to increase the number of bits to 16, 32 or (as in the case of the Intel Pentium and Digital Alpha) to 64.

The number of lines in the address bus limits the number of different addresses within the system (see table, p. 30). The Z80 and many other microprocessors have an address bus that is 16 bits wide. This allows it 64K addresses, which is enough for a small system. More powerful processors have more address lines, such as the 68000 family with 24 bits, and the Pentium with 32 bits.

LSB: least significant bit.

MSB: most significant bit.

Integer: a 'whole number', one without fractions. *Examples*: 145, 8, 0, –72.

disconnected from the internal circuit. The logical state of the internal circuit can then have no effect on the bus.

Whether the data pins of a device are in the high impedance state or not is decided by the level applied to its $\overline{\text{CHIP ENABLE}}$ ($\overline{\text{CE}}$) input. If the chip is enabled ($\overline{\text{CE}}$ low) the outputs may be 0 or 1, that is the voltage at each output pin is either 0 V or 5 V. The device puts this data on to the bus. If the chip is disabled ($\overline{\text{CE}}$ high) the pin is in the high impedance state so it can not place data on the bus. A control signal, sent by the CPU to the $\overline{\text{CE}}$ input of the device ensures that the device puts data on the bus only when required. At other times, it is disconnected. Only one device must be enabled at any one time. When one is enabled, the others must be disabled, otherwise two or more devices will try to put signals on the bus at the same time. This is known as *bus contention.*

What must the CPU do before it puts data on the data bus?

The $\overline{\text{CHIP ENABLE}}$ pin is sometimes known as the $\overline{\text{CHIP SELECT}}$ pin.

A line over the name or symbol of an input pin indicates that the action occurs when the input is made low. It is an *active-low* input.

The address bus and address decoding

Only the CPU puts addresses on the address bus. The other devices in the system simply wait until they are addressed before doing anything.

A device recognises when it is being addressed (that is, when its address is on the address bus) by means of a *decoder* circuit. This is a logic circuit that produces a low (usually) output when, and only when, its own address is on the bus. Fig. 2.10 illustrates the principle of address decoding with a practical circuit. The circuit is practical in the sense that it uses obtainable types of logic gate. An address consists of a set of 0's and 1's (lows and highs). The decoder must respond only when:

- all the address lines that are supposed to be high are high, and
- all the address lines that are supposed to be low are low.

In Fig. 2.10 the highs are taken care of by the 13-input NAND gate (the 74133). For the address in the example, the address lines that are

Memory size

Every location in the memory of a computer must be identified by its own unique *address.* The number of different addresses that are possible in a given system depends on the number of bits it its addresses. For example, many microprocessors use 16-bit addresses. The addresses run from 0000000000000000 (zero decimal) to 1111111111111111 (65535 in decimal), giving a total of 65536 possible addresses. With such large numbers, we usually quote them in 'K'. Really, K stands for 'kilo' which means 'thousand' but in quoting addresses and sizes of memories it means 'times 1024'. In the example, 65536 is quoted as 64K.

Here are some sizes of whole memories or of individual memory chips:

Bits	Addresses	In K . . .
8	256	–
10	1024	1K
12	4096	4K
14	16384	16K
16	65536	64K
18	262144	256K
20	1048576	1024K = 1M
24	16777216	16M
32	4294967296	4G
n	2^n	

The bottom line of the table explains why the numbers of addresses shoot up so rapidly. Do not try to remember these figures. Just remember the bottom line and you can always work out the rest on a calculator, if you ever need them.

Test your knowledge 2.4

A CPU has 13 address lines. How many locations can it address?

supposed to be high are connected to this gate. There are four spare inputs and these are wired to the positive supply line. When all inputs to the gate are high, its output goes low. We require a high output to send to the final gate (the 7410) and we use an INVERT gate to produce this. The 7400 family has no NOR gates with a large number of inputs so we use two 4-input NOR gates. The 7425 has two such gates on a single chip. The two gates handle the seven low lines in the target address. The eighth input is held low by connecting it to the 0 V supply line.

When the target address is on the bus, all three inputs to the 7410 go high. Its output goes low. This pin is wired to the $\overline{\text{CHIP ENABLE}}$ input of the device that is being addressed. This causes that one addressed device to become active.

There is more about addressing in Chapter 3.

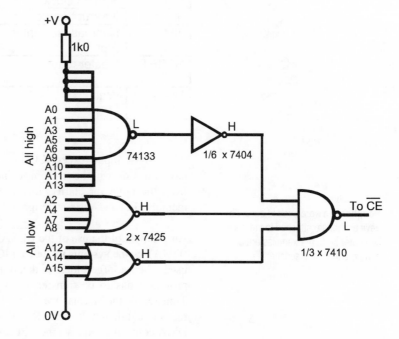

Figure 2.10 *It takes several logic gates to decode a single address. The diagram shows the logic levels present in the decoder when (and only when) the target binary address 0010 1110 0110 1011 (2D6C in hexadecimal) is present on the 16-bit address bus. The 1 kΩ resistor reduces flow of current to the unused inputs of the 74133 and so reduces the risk of damage from voltage spikes on the positive supply line.*

The control bus

The control bus consists of an assortment of lines with various functions. Some of them carry signals from the CPU to other devices. Some carry signals from other devices to the CPU. They *may* run side by side for part of their length, but their routing depends mainly on their function.

The composition of the control bus depends on the needs of the microprocessor. The Z80 control bus comprises these six input lines:

Symbol	Function	Details
BUSRQ	Bus request	Forces CPU to let device use the bus
CLK	System clock	Controls timing of cycles
INT0	Interrupt request 0	Device interrupts CPU
NMI	Nonmaskable interrupt	Interrupt of higher priority than INT0
RESET	Master reset	Initialises CPU and other devices
WAIT	Wait	Devices extend bus cycle to more than 1 clock period

All of these inputs except CLK are active low. Their symbols should have lines drawn above them to indicate this, but they were omitted from the table for clarity. In this table, the term 'devices' includes memory, I/O and various peripheral devices.

Peripheral: a word used to describe devices that are attached to the main computer system. Examples are a printer, a mouse, a scanner.

An example of the use of the $\overline{\text{BUSAK}}$ input is in *direct memory access* (DMA). Some systems include an IC known as a *DMA controller*. It is used when a large block of data (such as a word-processing file or a program) has to be transferred to memory from storage on a disk. Transfer in the normal way, in which the CPU requests each byte, reads it and then writes it to RAM, would take far too long. Instead the DMA controller requests the CPU to give up control of the address and data busses while it transfers the data directly from the disk to memory.

In this CPU, there are only two levels of interrupt, but other microprocessors may have more. A low input to NMI can not be ignored by the CPU unless a BUSRQ signal has been received. On receiving an NMI signal, the CPU completes the instruction it is working on and then jumps to the address of the interrupt request routine (IRQ). The INT0 signal is the one normally used for requesting interrupts. This can be ignored if the interrupt flag in the status register has not been set. This allows the programmer to disable interrupts when the CPU is engaged in a complicated routine that might crash if interrupted. The I flag can be set after the routine is complete to enable interrupts again.

The seven output pins of the Z80 control bus are all 3-state outputs:

Symbol	Function	Details
BUSACK	Bus acknowledge	Responds to BUSREQ by making this low and making its data and address outputs high impedance
HALT	Halt	The Z80 has stopped because of a HALT instruction
INSTRD (or M1)	Instruction read	The Z80 is reading an instruction from memory (MREQ and RD also low)
IORQ	I/O request	The Z80 is reading or writing data to I/O
MRQ	Memory request	The Z80 is reading or writing data to memory
RD	Reading data	Causes device to put data on bus
RFSH	Refresh	Signal to refresh dynamic memory
WR	Writing data	The data bus has data ready for a device to copy

All of these outputs are active low. Their symbols should have a line drawn above them to indicate this, but they were omitted from the table for clarity.

When the Z80 is about to read or write to memory it makes the $\overline{MRQ}$ line low. This is the equivalent of a CHIP ENABLE signal to the memory chip. At the same time, it makes the $\overline{RD}$ or $\overline{WR}$ lines low, depending on whether it wants to read or write data. Similarly, when it wants to read or write to a port it makes the $\overline{IORQ}$ line low, and makes the $\overline{RD}$ or $\overline{WR}$ low at the same time.

The Z80 has separate lines for enabling reading or writing. Some other CPUs, such as the 6205, use a single line, symbol $R/\overline{W}$. A memory IC interprets this as a read operation when it is high and a write operation when it is low.

On studying the two tables above, it can be seen that the majority of the control lines are used for handshaking between the CPU and other devices. In other words, the control lines are the means by which the actions of the CPU and of the other devices are co-ordinated.

Other processors – the 6502

Discussion of microprocessors in this chapter is centred mainly on the Zilog Z80, but there are many other processors in use. They have individual features, which may make them better suited for certain applications.

The Synertek 6502 is a relatively early processor that has been very successful in the past and is still in use today. Its architecture has the basic features (internal busses, status register, stack register, program counter, address and data buffers, ALU) that are in the Z80. It differs in having only an accumulator and no general purpose registers. This means that all processing centres on the accumulator and ALU, a difference that shows in the types of instructions in its instruction set.

Like the Z80, it has an 8-bit bi-directional data bus and a 16-bit address bus. It has two interrupt inputs, NMI and IRQ, which function in a similar way to those in the Z80.

The 6502 has a wide range of addressing modes, some of which help to make programming simpler. One of these is *zero page addressing,* which assumes that the first (high) byte of an address is $00. Instead of quoting the full address, such as $005A, the programmer quotes only $5A. This shorter form saves program storage space and runs faster. The 6502 makes use of its two *index registers,* X and Y for addressing. In *indexed zero page addressing,* an address in zero page is given relative to a value stored in the X or Y register. For example if the X register holds $24 and the address is given as $57, the actual address is found by adding these together to give $007B. This feature has the advantage that the registers can have different values put into them as the program runs. If the program runs in a loop, X can be changed each time round the loop so that different addresses are accessed each time. In particular, X and Y can be incremented or decremented so the program can scan a table of data stored in consecutive bytes in zero page. Indexed addressing is not restricted to zero page. Other modes of addressing apply it to addresses in the whole memory space.

System clock

The clock is a squarewave generator capable of running at high frequency. Some microprocessors (such as the 8085) and practically all microcontrollers include the clock circuitry on the chip so all that is necessary is to attach a crystal having the required frequency. If a completely separate clock is required, this can be obtained as an IC, such as the 6872. With a suitable crystal, this can operate at frequencies up to 10 MHz.

Another source of clock pulses is a crystal oscillator module, which includes the crystal and logic circuitry required to give a TTL output. These can be obtained to run at frequencies up to 50 MHz, with suitable short rise-time and fall-time. In many systems, a frequency of 1MHz is fast enough, in which case it is feasible to assemble a clock based on standard logic gates (Fig. 2.11).

In PCs and other computing systems, processors are clocked to operate very rapidly so as to process as much data as possible in a given time. This becomes very important when computers are running graphics and multimedia programs. It is also important when clocking telecommunications systems, so that data may be transferred as rapidly as possible. Clock frequencies of several hundred megahertz are common and frequencies of 1 GHz or more are attainable.

The frequency quoted for a CPU is usually the maximum frequency at which it can reliably be run. It can be run at frequencies less than the maximum because the clock pulses co-ordinate all parts of the system to operate as one.

The *system* clock must not be confused with the *real-time* clock, which is described in Chapter 4.

> **Test your knowledge 2.7**
>
> If the system clock is running at 2 MHz, what is the length of a single cycle?

GHz: gigahertz, which is equal to 1000 MHz.

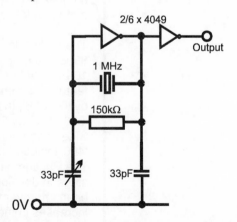

Figure 2.11 *A system clock built from two CMOS INVERT gates. The gate on the right is a buffer to avoid over-loading the oscillator and so alter its frequency. This clock has a 1 MHz quartz crystal but crystals of other frequencies can be used.*

Other processors – the 8086 family

When we say that the Intel 8086 is a 16-bit processor, we mean that its data bus is 16 bits wide. Instead of fetching data a byte at a time like the Z80, 6502 and many other processors do, it fetches a double byte or word. This gives it a big increase in speed of operation. Another innovation that increases speed is the *prefetch buffer* in which instructions are queued, ready for execution. The address bus is 20 bits wide, so the 8086 addresses up to 1 Mb of memory. To increase the width of the buses while still keeping the processor in the standard 40-pin package as used by the Z80 and 6502, some of the pins have been *multiplexed*. They perform different functions at different stages of the operating cycle. Fifteen pins (AD0 to AD14) are used both for the data bus and for the lower fifteen lines of the address bus. An example of this kind of multiplexing appears in Chapter 8. The four upper pins (A16 to A19) of the address bus are multiplexed as status output pins. One of the control pins is the MN/$\overline{\text{MX}}$, which selects between two modes of operation. When this is high, the processor is in *min mode*; it operates much like the earlier 8085 processor. When the pin is low, it operates in *max mode*; it is able to work in conjunction with other processing ICs such as a *maths coprocessor*, which is specially designed to take over the more complex mathematical operations of the processor. This gives an advantage of speed when running programs in which mathematical operations predominate.

The 80286 (known as the '286') was the successor to the 8086 in a 68-pin package with a 24-bit address bus and a 16-bit data bus. It can access up to 16K of memory. This featured a number of improvements and was succeeded by the '386', the '486' and the Pentium processors. The '386' is a true 32-bit processor, with all its registers 32 bits wide. It can access 4 Gb of memory. Some versions of the '486' included the maths coprocessor on the same chip. The Pentium includes this as a regular feature. At every stage in development from '286' to Pentium there have been improvements in performance. There have been increases in the number of transistors (now over a million), the number of pins, the number of instructions in the set, and the maximum clocking rate.

Activity 2.1 Other processors

This chapter describes some of the features of the '1200', the PICs, the Z80, the 6502 and the 8086 family. Select one or two other popular processors from the same or different families and briefly study the manufacturer's data sheets. Write a short account of the main features of each processor and describe their advantages or usefulness.

Activity 2.2 Address decoding

An imaginary microprocessor has an 8-bit address bus. Design a decoder that responds when the binary address on the bus is 1100 0101. Build the decoder on a breadboard, using CMOS or TTL ICs, and test its action.

Logic families

There are three main logic families:

- Transistor-transistor logic (TTL).
- Complementary MOSFET logic (CMOS).
- Emitter coupled logic (ECL).

Each of these families has its own versions of the standard logic gates, such as NAND and NOR. Each family includes a range of ICs with more complex functions such as flip-flops, adders, and counters. The families differ in the way the basic gates are built, and this gives rise to family differences in operating conditions and performance.

Fig. 2.12 shows the logic levels used by TTL and CMOS families.

The main features of each family are as follows:

TTL: Based on bipolar transistors. Operates on 5 V DC, which must be regulated to within ±0.25 V. The original 74XX series (all type numbers begin with '74') is almost completely replaced by newer series with improved performance. One of the most popular is the 74LSXX series, which has a lower power requirement of about 2 mW per gate. It has faster operation than standard TTL, the typical propagation time per gate being 9 ns. It operates with clock speeds up

---

Bit slicing techniques

While most systems are based on processors with a data bus width of 8 bits or more, another approach uses 4-bit controllers. Such a system may have 2, 4, or more controllers working in parallel. To process a 16-bit value, for example, the data bits may be divided into four 'slices', each four bits wide. The slices are processed simultaneously and the results combined into a 16-bit word. The advantage of this technique is that the 4-bit processors are faster than larger processors and, working together at the same time, produce the result much more quickly.

---

to 40 MHz. Some of the newer series operate on 3.3 V, making them suitable for battery-powered equipment. TTL is widely used in microelectronic systems.

CMOS: Based on MOSFETs, and therefore have very low power requirements, typically 0.6 mW per gate. Operating voltage is in the range 3 V to 15 V (absolute maximum = 18 V), which makes it ideal for battery-powered equipment. With FET inputs, the gates require so little input current that the output from a gate can be fanned out to 50 or more gates. The most popular series is the 4000B series. Its main limitation is speed: its propagation time is in the region of 125 ns and the maximum clock rate is 5 MHz. This may not matter in many

Test your knowledge 2.8

What is the margin of error on a logic low output from a TTL gate?

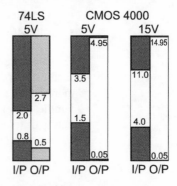

Figure 2.12 *Voltage levels for TTL and CMOS are specified as shown in these diagrams. For example, with a 5 V supply a CMOS gate interprets an input below 1.5 V as a logic low input. An input above 3.5 V is taken to be a logic high input. A low output is always between 0V and 0.05 V, and a high output is between 4.95 V and 5 V. In either case there is a 1.45 V margin for error. A 'noise' spike on an output signal can be as much as 1.45 V without affecting its apparent logical value. This means that CMOS has good noise immunity.*

microelectronic applications. Several of the 74XX series are available in CMOS versions, known as the 74HCXX and 74HCTXX series. Logic 0 and logic 1 voltage levels are widely separated (see Fig. 2.12) making the series highly immune to noise. Because of the small size of CMOS gates, it is possible to fabricate complicated logic circuits on the chip. The series includes many complex devices, such as the 4020 14-stage counter, which are not available in the 74XX series.

ECL: This is specially designed for high-speed operation. The transistors are never driven into saturation so they quickly change state when logic inputs change. In addition, the circuits have low input impedances to avoid the speed-reducing effects of capacitance. This increases power requirements to around 30 mW per gate. Propagation time is about 1 ns per gate and the maximum clock rate is in the region of 500 MHz. Because the transistors are operated in the non-saturated state, logic levels have to be carefully controlled at -0.8 V and -1.6 V respectively. The small difference between these levels means that ECL is much more affected than the other families by noise spikes on the lines. Specially filtered power supplies are essential. Because of the high frequency of the signals, tracks on the circuit board have to be much more carefully laid out. In summary, ECL is difficult to use and is restricted to computer circuits where very high speed is the prime factor.

Problems on the CPU

1 Describe the main features of a named microcontroller (details of internal structure not required).

2 Explain the difference between parallel and serial transfer of data. What are the advantages and disadvantages of each?

3 Describe the main features of a named microprocessor (details of internal structure are not required).

4 Name the three buses of a microprocessor system. Outline their features and functions, giving examples.

5 What are three-state outputs and why are they essential on the data bus? Why are they not needed on the address bus?

6 Explain, giving an example, what is meant by *address decoding*.

7 Describe the signals present on the control bus when a Z80 (or other named microprocessor) is writing data into memory.

8 What is a system clock and why is it an essential unit in a microelectronic system?

9 Explain what bit, nybble, byte and word mean.

10 What is meant by bus contention and how may it be avoided?

Multiple choice questions

1 When it is essential for a Z80 to be interrupted, a low signal is put on:

    A the data bus.
    B the NMI line.
    C the control bus.
    D the INT0 line.

2 Signals can be sent in either direction on:

    A the address bus.
    B the BUSACK line.
    C the WAIT line.
    D the data bus.

3 Three-state outputs are used:

    A for putting addresses on the address bus.
    B to send interrupts.
    C for reading data from the data bus.
    D to isolate internal circuits from the data bus.

4 A system has an address bus that is 14 bits wide. The most significant bit is also known as:

    A A14.
    B A0.
    C A13.
    D A1.

5 An example of an integer is:

    A -24.
    B π.
    C 3.76
    D 38½.

6 One of the logic families most suited for high speed operation is:

    A ECL.
    B TTL
    C 74LS00.
    D CMOS.

7 A logic family most suited for battery-powered equipment is:

    A TTL.
    B ECL.
    C CMOS.
    D 74LS00.

3　Memory

Summary

The two main types of memory are random access memory (RAM) and read only memory (ROM). Two types of RAM are static RAM (SRAM) and dynamic RAM (DRAM). There are several types of ROM, including mask-programmed ROM, PROM, EPROM and EEROM. Decoders are used to route signals to or from a specified location in memory. A real time clock relieves the CPU of timekeeping duties.

There are two main types of memory:

- Random access memory – used for temporary storage. Data may be written into it at any time, and later read from it The data is lost when the power supply is switched off. Used for the storage of data and for programs copied from more permanent data stores such as magnetic disks.

- Read only memory – used for permanent storage. In most types, the data once written into it can not be changed. Used for storing programs and data tables.

Random access memory

This occupies most of the available address space in a typical microcomputer. A PC when purchased may be equipped with 32 Mb of RAM, with room for expansion to 64 Mb or more. The RAM is used for the temporary storage of programs and data. There is usually less need for RAM in a microcontroller system, as their programs are permanent and are stored in ROM instead (see below). The RAM is

Mb: the symbol for *megabyte*, which is approximately one million bytes, but more exactly is 2^{20}, or 1048576 bytes.

usually included on the microcontroller chip, and may consist of as few as 64 *bytes*.

Whenever data is read from RAM, it is *copied* in a register of the CPU. Conversely, it is copied from the CPU when it is written into RAM. In either case, the original data remains unaltered after copying. This is known as *non-destructive readout*.

Random access: data may be read from or written to any location at any time. This is different from, say, a shift register, from which data must be read in the order in which it was written.

Once data is stored in RAM, it remains there until it is either:

- Overwritten and replaced by new data, or
- The power supply is switched off.

There are two types of RAM:

K and **Kb**: the K symbol indicates approximately one thousand, but is more exactly 2^{10}, or 1024. The Kb is a kilobyte or 1024 bytes.

Static RAM (SRAM). Each SRAM IC contains an array of flip-flops, each one of which can be set or reset to one of its two states and so represent a logic 0 or a logic 1. The flip-flops may be individually addressable, or they may be addressed in groups of 4 (a *nybble*) or 8 (a *byte*) or 16 (a *double byte*, also known as a *word*). An example is the 6116 SRAM, which has its 16K flip-flops, arranged in 2K groups of eight (Fig. 3.1). The address decoder is included on the chip. To address 2K locations requires 11 address lines (2^{11} = 2048). Since each location stores a byte, the IC has eight data lines for input/output. Typically, the time needed to read or write a byte of data (the access time) is 100 ns compared with 20 ns for a hard disk and 200 ns for a floppy disk.

Flip-flop: a circuit based on two transistors, connected so that when one is on the other is off. The circuit is stable in either of two states, which can be made to correspond to logical 0 and 1.

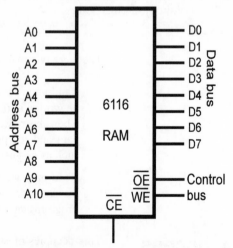

Figure 3.1 *The 6116 SRAM holds 2 KB bytes of data. It therefore needs 11 address inputs and 8 data inputs/outputs. It has OUTPUT ENABLE and WRITE ENABLE inputs which must be made low for reading and writing respectively, but the chip will take no action unless the CHIP ENABLE input is made low at the same time.*

Test your knowledge 3.2

What is the feature of MOSFETs that makes them able to store electric charge?

Dynamic RAM (DRAM). A DRAM IC consists of an array of MOS-FET transistors. Each of these can be switched on by storing a charge on its gate. Writing to DRAM consists of charging or discharging the gate. Reading consists of finding out if the transistor is on or off by registering the level (0 or 1) it produces on the data line. The single transistor of a DRAM location is much smaller than the flip-flop of a SRAM location so many more locations can be accommodated on a DRAM chip. For example, a typical large SRAM IC stores 4 Mb, but the equivalent large DRAM stores 16 Mb. This makes large memories easier to build. In addition, DRAM access time is only 70 ns, which is an advantage in computers, for high-speed operation.

The problem with storing a charge is that it eventually leaks away. This means that the data stored in the transistor must be refreshed at regular intervals. Under the control of a clock running at about 10 kHz, the remaining charge on a transistor is passed on from one transistor to another and topped up during the transfer. In this way the data is kept 'on the move', giving this type of memory the description 'dynamic'. The need to refresh memory means that a certain amount of the computer's operating time must be set aside for this. The DRAM can not be used for reading or writing during this time. This makes the operating system more complicated to run.

Read only memory

ROM is a permanent or semi-permanent form of memory. Unlike RAM, it does not lose its stored data when the power is switched off. This is why it used for storage of the booting routines in microprocessor systems. In microcontroller systems such as the cordless telephone and the multimeter, the program is unalterable and is stored in ROM during manufacture. In the larger more 'intelligent' microcontroller systems such as a data logger there is ROM to store the basic routines, with a medium-sized RAM as a temporary store for programs that are currently in use.

There are several kinds of ROM, including:

Mask-programmed ROM: Special masks are used when the chip is made, so that the content of every memory location is fixed from the beginning. It can never be altered. In other words, the ROM is a special kind of logic device in which each different combination of logic levels on the address bus produces a corresponding combination of logic levels on the data bus. Making special masks is very expensive, so this type of ROM is used only when it is known in advance that thousands of them will be required. For example, it might be suitable to mask-program a ROM for use in a popular model of a washing machine.

PROM: This is programmable ROM, which is programmed *after* it is manufactured, but the program can not then be altered. It is sometimes referred to as *one time programmable* (OTP) ROM. Fig. 3.2 shows how the programming is done. In the simplified circuit in the diagram, there are only three address inputs for the memory locations, which can be decoded to one of eight addresses. There are four data output lines, which run across the lines of the memory locations. At each place where the two sets of lines cross there is a transistor connected as in Fig. 3.3. The emitter of this has a fusible link connected to it. This thin connection can be 'blown' by passing a relatively large current through it when the ROM is being programmed. To blow it, we connect the data output line to 0 V and then apply a high level to the memory location line. If V+ is made higher than usual, a large current flows through the transistor and melts, or fuses, the link. All of this can be done automatically with the PROM plugged into a socket on a special PROM programmer. This is loaded with the data or program to be put into the PROM. It works through the memory locations one at a time, either blowing a link or leaving it unblown.

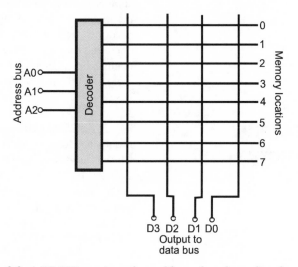

Figure 3.2 *A PROM consists of an address decoder, a line for each memory location and a set of data output lines. Where they cross, there may be a link between the memory line and the data output line.*

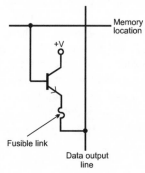

Figure 3.3 *There is a transistor at every crossing point in Fig. 3.2, linking the memory location to the data line through a fusible link. This link may be blown when the PROM is programmed.*

Example:

Fig 3.4 shows the links of memory location at address 011 (3 in decimal). Two have been blown and two left unblown. After the PROM has been programmed, it is ready for use as ROM in a microelectronic system. The voltage +V used for reading the ROM is not as high as it was for programming. Turning on the transistor now does not blow the unblown links. When a memory line is addressed, it goes high and turns on the transistors that have unblown links. Currents flow through these to the data bus and are equivalent to a stored '1'. No current flows where a link is blown and is equivalent to a stored '0'. In Fig 3.4, the blown and unblown links produce a nybble of data equivalent to 1010.

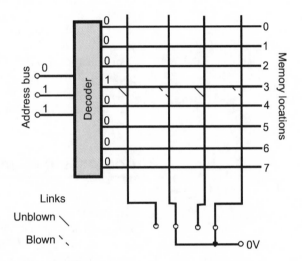

Figure 3.4 *The PROM is being programmed with data 1010 in locations 011. An extra high voltage is applied to the location line, while the appropriate data lines (D0 and D2) are connected to 0 V. This blows the two links where shown, leaving the other links intact. When the PROM is in use, a high level on the location line produces high levels on D1 and D3, but D0 and D2 remain at logic low.*

Blowing the links destroys them permanently, so a PROM can be programmed only once. If a mistake is made in programming, or the data needs to be altered, another PROM must be programmed with the corrected data. In the early stages of development of a system PROMs are programmed singly until the program and the device in which it is to be used have been fully tested. Once a correct version of the program has been developed, the mass programming process is automatic and reasonably inexpensive.

EPROM: This is *erasable programmable ROM*. The ROM consists of an array of MOSFETs, one for each bit. During programming, a charge is pulsed on to the gates of those MOSFETs that are to be set to '1'. The principle is similar to that of DRAM except that the charge remains for very much longer, usually for several years. EPROMs have a quartz window though which the chip can be exposed to ultraviolet light if the data is to be erased. Ultraviolet is an ionising radiation and the ions produced in the chip discharge the gates and erase the data.

An EPROM can be reprogrammed after erasing, which makes it very suitable for use when a system is at the development stage and frequent revision of the program and data are needed.

EEROM: This is *electrically erasable ROM*, which can be programmed and reprogrammed while still connected in its circuit. The ROM of many microcontrollers is of this kind. Although this type of ROM could be used for temporary storage in a way similar to RAM, writing new data into EEROM is too slow for this to be practicable. Some microcontrollers are made in two versions, with an EEROM program memory or with a PROM program memory. The EEROM version is used at the development stage, then the PROM version is used for mass-production of the finalised program.

Addressing memory

The example given in Fig. 2.10 explained how to decode a single device located at one particular address in the computer's address space. In practice, devices such as disk drives, I/O ports and the keyboard have several registers in them and each of these has its own address. Fig. 2.10 shows several gates being used to decode just one address. However, it is possible to decode a number of adjacent addresses with only a small amount of additional logic. Devices which occupy a range of addresses often have a decoder included on the same chip, so there are fewer problems for the system designer.

A memory chip is a good example of a device which occupies a block of adjacent addresses, and which contain its own decoder. In Fig. 3.5, the 6116 RAM chip has 2K (2048) memory locations within it. Each location has 8 memory cells, so each location holds a byte of data. The chip has 8 data input/output pins so that the microprocessor can either write or read a byte at a time. Whether there is a read or a write operation depends on the state of the two control lines shown in Fig. 3.5. If the OUTPUT ENABLE is made low, the addressed data is put on the bus. If the WRITE ENABLE is made low, data from the bus is stored on the chip at the addressed location.

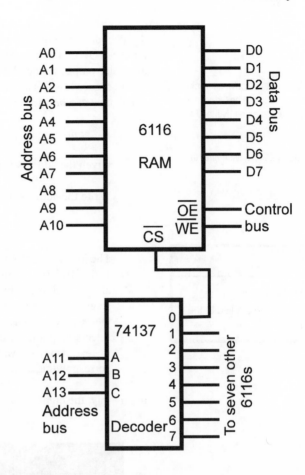

Figure 3.5 *Decoding the 16K addresses within the eight 6116 memory ICs is made simpler because each 6116 has its own internal decoder and the 74137 contains all the logic needed to decode the upper three address lines.*

The 2K storage locations of a 6116 require an address bus 11 bits wide to address every location. The 6116 has a built-in decoder with 11 address inputs A0 to A10. To increase the amount of memory in a system we can use several 6116s, and access these one at a time. Each 6116 is connected to the data bus, to the control bus, and to bits A0 to A10 of the address bus. The eleven address bits address the corresponding location in every one of the eight chips.

The next problem is to select just one of the eight chips. To select one out of eight requires three lines, which we will name A11, A12 and A13. The figure shows how we use them. They go to a decoder IC known as a 3-line-to-8-line decoder (or multiplexer). The three address lines are connected to the input lines A to C. Normally, the outputs of the decoder are high but, when any *one* output is addressed by the inputs at A to C, it goes low. When all three inputs are low for example, output 0 goes low. When A and C are high and B is low, output 5 goes low. As an output goes low it enables the selected 6116, which then can be written to or read from.

Example:

Given the address $1E6D, how is this decoded?

Write the address in binary:

0001 1110 0110 1011

The lower 11 bits go directly to the 6116s. The lower 11 bits are 110 0110 1011 in binary or $66D, so byte $66D is addressed in all eight chips. Which of the eight chips will function at this stage depends on the top 3 bits of the address, which are 011. In the decoder, input A is 1, input B is 1 and input C is 0. This selects output 6, so the *seventh* chip in the set is selected. Data is read from or written to this chip only. The data at the corresponding address in the other seven chips is ignored.

Test your knowledge 1.1

Describe the decoding of the address $2B72 if applied to the circuit of Fig. 3.5.

In this way, the eight chips cover these addresses:

Chip no.	Address range (in hex)
0	0000 to 07FF
1	0800 to 0FFF
2	1000 to 17FF
3	1800 to 1FFF
4	2000 to 27FF
5	2800 to 2FFF
6	3000 to 37FF
7	3800 to 3FFF

The highest address is $3FFF. In decimal, this is 16383, so the set of eight chips stores 16K. Just to confirm the calculation, the address bus has 14 (11 + 3) bits, so the number of addresses is $2^{14} = 16384 = 16K$.

Many systems have a 16-bit address bus, but the circuit of Fig. 3.5 is unaffected by the values of the top two bits. These can take the values 00, 01, 10 or 11, but they are not decoded and make no difference to the result. The effect of this is that there are 'ghost' addresses with different values for the top two bits.

Example:

Consider the lowest address in chip no. 5, which is $2800. Preceding this address by values of lines A14 and A15 can alter the value of the most significant hex digit. The last two bits of this remain the same (10) but the first two may vary. We can have 0010, 0101, 1010 or 1110. In hex these are 2, 6, A and E. The real address is $2800, and the ghost addresses are $6800, $A800 and $E800. Chip no. 5 is enabled if any one of these ghost addresses is placed on the bus.

Allocation of addresses

With a 16-bit address bus, for example, a system has 64K different addresses. This is the *address space* of the system. This does not mean that there is a device or memory location at every address, waiting for the CPU to call it into action. The address space is the range of *possible* addresses. The addresses that are actually used depend on what devices are present in the system. When you purchase a computer, it may often have only a limited amount of memory. Later you may want to expand the system by purchasing additional memory boards. These plug into sockets on the motherboard, which are already provided with decoders to place the new memory in a range of previously unused addresses. The same applies with new input/output devices that you may wish to add to the system. Each must be allocated its individual address, or range of addresses. Often the operating system takes care of this automatically, but there can be occasions on which two different devices are allocated the same address. This always leads to problems of conflict.

The allocation of addresses may be visualised by drawing an *address map*. This is often called a *memory map* because most of it is occupied by memory of various kinds. The map in Fig. 3.6 is typical of a small computer or microprocessor system with a 16-bit address bus, capable of addressing up to 64 kilobytes. The first kilobyte is given over to one or more memory chips that store the booting routines. When we start (or *boot*) a system the microprocessor needs to be told how to get the system ready for operation. It automatically goes to address $0000 and there reads the first byte of the first instruction. Reading on through the bottom kilobyte it finds further instructions for getting the system *booted up*. This first kilobyte also contains interrupt service routines. The purpose of these is explained later. Note that the routines in the first kilobyte must be available when the system is first switched on. For this reason, the memory in this range of addresses is ROM.

Usually the input/output devices such as the printer and the disk drives

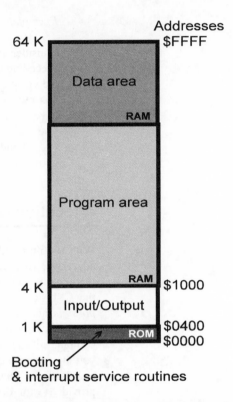

Figure 3.6 *The address allocations of a typical microelectronic system that has a 16-bit address bus.*

are located next above the booting and ISR routines. There are spare addresses in this space, so that other devices can be added to the system when required.

The upper part of the address space is occupied wholly or partly by RAM. The area is usually divided into two parts, the program area and the data area. The program area is used for holding the program on which the system is currently working. With many complex programs, there is not enough address space to hold the whole program at once. Different sections of the program are loaded into the program area from the disk drive and replaced with other sections as the microprocessor requires them. This is why you often hear the disk drive of a computer switching on and off automatically while you are running a program. The top end of memory is generally used as a data storage area. Here it stores data on which it is currently working, including values obtained in the intermediate stages of calculations. Data is being stored, read, and replaced continually as the system goes about its tasks.

A system can operate with only a portion of its program and data areas actually occupied with memory. The system adjusts itself to work within whatever memory is available. The larger the memory, the larger the sections of a program that can be held there at one time. The larger the sections, the less often the microprocessor has to pause in its operations while the next section and its data are loaded.

Real time clock

Every system must have a *system* clock to drive the processor. A real time clock is optional. It functions like a digital watch, keeping account not only of the time but also the date, the day of the week, the day of the month, the month and the year. It includes logic to make it give February twenty-nine days in a Leap Year. It also has an alarm function to produce a signal at any preset time. In addition, it can be programmed to generate a pulse to interrupt the CPU at regular intervals.

To the CPU the real time clock is a block of addresses in which time and date are updated automatically once they have been set. The CPU writes into or reads from these addresses just as it does with memory. In some types of RTC the addresses used for times and dates are part of a larger block of RAM that is available for other purposes. The RTC always has to have its own back up battery so that it does not stop working when the system is switched off. This makes the clock's spare RAM useful for storing data that must not be lost.

All of the functions of the real time clock except interrupt generation could be performed by the processor itself, if it was programmed to do so. This would have two serious disadvantages. One is that the processor would be kept so busy with time-keeping functions that it might have too little time to spare for doing anything else. Another disadvantage is that the action of a processor can be interrupted when it has to attend to something urgent. Interrupting the clock action would make the clock lose time. It is better to give the time-keeping task to an independent IC.

There is a detailed study of an RTC in Chapter 10.

Activity 3.1 Investigating memory

Connect up a 6116 or another DRAM chip on a breadboard. It needs connections to the positive and ground (0 V) rails of a +5 V supply. Input to the address, data and control terminals of the IC can be provided through switches (Fig. 2.6) or by simply plugging in wires to connect them directly to the supply line or 0 V line.

Output from the data terminals may be read by connecting LEDs, as in Fig. 2.7. Either use a low-voltage filament lamp as shown or LEDs in series with a 270 Ω resistor. The base resistor of each transistor should be about 10 kΩ.

Try to store a value in one of the locations of the IC. Make the control inputs high so that they are inactive. Set up the address bus and the data bus with suitable values. Then make the write enable line low. Finally, make the chip enable line low, then high again to latch the data. Note that these instructions may need to be modified to suit the type of RAM used.

Now try to read the data you have written, with LEDs connected to the data outputs. Set up the address but to the same address as before. Make the output enable line low, then the chip enable line.

Activity 3.2 Addressing memory

Extend the circuit used in Activity 2.1 by adding a decoder, as in Fig 3.5. Decide on one or two addresses, set the lower bits by connections to the RAM. Set the upper bits by connections to the decoder. Connect the chip enable input of the RAM to the appropriate output pin of the decoder.

Try to write and read to various addresses in the RAM.

Problems on memory

1 Describe RAM and the way that it is used in microelectronic systems.

2 Compare the functions of RAM and ROM.

3 List the different kinds of ROM, and give examples to illustrate the ways in which each kind is used.

4 Draw a diagram similar to Fig. 3.4 but with 4 address lines and 6 data lines. Show the state of the links when the value decimal 35 has been blown into address $5, and the value decimal 21 has been blown into address $E.

5 On a copy of Fig 3.5, write in the voltage levels (0 or 1), that would be present when the value decimal 99 is being written into address $0372.

6 Design a SRAM to store 128 Kb (kilobytes) of data in four blocks of 32 Kb each. Use a supplier's catalogue and/or manufacturer's data sheets to select a suitable type of RAM and decoder. Draw a diagram to show all the connections.

7 Repeat question 6 for a SRAM to store 48 Kbits of data in three blocks of 16 Kbits each.

Multiple choice questions

1 Which type of memory must be refreshed while the system is running?

    A EEROM.
    B SRAM.
    C DRAM.
    D Mask-programmed ROM.

2 Which type of memory is most suited to store the program in a reprogrammable microcontroller?

    A EEROM.
    B DRAM.
    C PROM.
    D Mask-programmed ROM.

3 How many address lines are needed to address all the locations in a 4K ROM?

    A 10.
    B 12.
    C 14.
    D 17.

4 How many locations can be addressed by a CPU with a 14-bit address bus?

    A 16 384.
    B 8192.
    C 8191.
    D 32 768.

5 The number of address lines of a ROM that stores 1 Mbits as bytes is:

    A 16.
    B 20.
    C 17.
    D 10.

6 To make a memory chip store data from the bus we must:

    A make the $\overline{\text{WE}}$ line and $\overline{\text{CE}}$ lines low.
    B make the $\overline{\text{OE}}$ line low.
    C make the $\overline{\text{WE}}$ line high.
    D make the $\overline{\text{CE}}$ line low.

7 To make a memory chip place stored data on the bus we must:

A make the $\overline{CE}$ line high.

B make the $\overline{WE}$ and $\overline{CE}$ lines low.

C make the $\overline{CE}$ and $\overline{OE}$ lines low.

D make the $\overline{OE}$ and $\overline{WE}$ lines low.

8 The typical access time of DRAM is:

A 70 ns.

B 100 ns.

C 25 ms.

D 250 ns.

9 The typical access time of SRAM is:

A 70 ns.

B 10 ns.

C 250 ns.

D 1 ms.

10 A kilobyte is:

A 1000 bytes.

B 1024 bits.

C 1024 bytes.

D 1000 bits.

11 The kind of memory used in a microcomputer for storing the booting up program is:

A cache.

B SRAM.

C DRAM.

D ROM.

4 Inside the CPU

Summary

The criteria for the choice of a CPU are briefly set out. After an introduction to the nature of a program, the functions of the units within the CPU are discussed in detail. Measures to increase operating speeds are discussed.

When we are choosing a CPU for a microelectronic system, there are a number of points to consider:

- **Microprocessor or microcontroller?** If we are building a system that is required to process a large amount of data very quickly, then we choose a microprocessor. We can back up its powerful features with plenty of memory and a wide range of peripherals and input/output devices. If high performance and complex data processing is not essential, and particularly if the system is to take up only a small amount of space and is to be inexpensive, or is an embedded system, then we chose a microcontroller.

- **Clock speed.** High speed may be essential for large-scale processing, especially real-time control and complex graphics. However, high speed makes the CPU expensive and may cause problems in the design of the circuit board. For many applications, a clock speed of 1 MHz is more than adequate.

- **Bus width.** The wider the address bus, the larger the address

Embedded system: when a micro-controller system is built as part of a larger system and not as a separate unit, we say is it embedded.

space. The wider the data bus, the easier and faster it is to process large numerical values more precisely. However, wide busses make it more difficult to lay out the circuit board. Wide busses are used in powerful computer systems, but an 8-bit data bus and a 16-bit address bus is sufficient for very many applications.

- **Instruction set.** The choice between a CISC processor (see opposite) and a RISC processor depends partly on the application. In general, RISC processors are considered to be the faster choice.

- **I/O facilities.** Microcontrollers differ widely in the number of I/O pins or ports.

- **On-chip facilities**. Microcontrollers vary in the amount of RAM (none or 25 bytes to 1 K), PROM (none to 8 K), EPROM/ EEPROM (none to 8 K), they have, and whether or not they have an ADC, a DAC, and one or more timers.

ADC: analogue to digital converter.

DAC: digital to analogue converter.

Although there are many different kinds of CPU, as the previous section made clear, there are many similarities between them. This is because, although some CPUs are faster than others, some address more memory than others, and some have a bigger instruction set than others, they all basically do the same thing. Before we go on to look inside the CPU, we will outline what it does.

Programs and processing

A computer program consists of a sequence of codes. For long-term storage these may be held on a disk (floppy disk, hard disk or compact disc) or tape. They are transferred to a block of RAM when the CPU is to read them. Alternatively, the codes are stored in ROM and read from there. Each code group is either:

- an instruction to the CPU, or
- a value for it to use.

Note the spelling of disk/disc.

For simplicity, assume that each code group consists of a single byte. This is actually the case in many CPUs but some read longer codes. If a code is 8 bits long, there can be 256 different codes, each corresponding with a different instruction. So there can be 256 different instructions to the CPU, which is more than enough for most of them. Similarly, there can be 256 different values (0, 1, 2,..., 255) which is certainly not enough for the calculations we may want the CPU to perform. However, there are ways of expressing larger numbers, negative numbers and also ways of expressing numbers with greater precision (for example numbers such as 23.456789). In Chapter 7, we will explain the ways of coding values. For the moment, we are thinking only of single-byte values.

A program written in a code that the machine (the CPU) can read and act on is said to be written in *machine code*.

The codes are not codes as we write or print them on paper. They are the states of sets of eight flip-flops (if SRAM) or eight MOSFET transistors (if DRAM), or eight fusible links (if PROM) and so on.

CISC and RISC

Performing a logical operation by using a circuit of logic gates (hardware) is about ten times faster than running the equivalent software on a CPU. Traditionally, manufacturers increased the speed of their processors by including extra logic gates connected to perform a wider range of operations. As a result, a larger number (several hundred) of different instructions is needed to operate the CPU. This type of organisation is known as a *complex instruction set computer*, shortened to CISC. Most of the 8-bit CPUs and more powerful ones such as the Intel Pentium and Motorola 68000, are CISC processors.

Designers have analysed the programs written for CISC machines and noted that, *in practice,* a relatively small number (20%) of instructions are used for the majority (80%) of operations. Other instructions are less often used. These instructions can be omitted (and also the internal hardware associated with them), leaving the MPU with *a reduced instruction set* (RISC) of fewer than a hundred instructions. It is also possible to eliminate the internal software that many CISC chips have for performing certain operations. This simplifies the structure and the operation of the MPU and makes it faster. It also has other advantages; for example, the instructions can be all the same length, making the processing faster. In many of the newer CPUs, all instructions can be performed in a single cycle of the system clock. Examples of RISC processors are the Digital Alpha, and the Intel 80960 microprocessors and the PIC17CXX microcontrollers.

If a RISC computer has to perform certain tasks that have been eliminated from the instruction set, it will have to be programmed to perform the operation in several steps instead of just one. It may take longer than a CISC processor. Because such operations are needed only rarely, there is an overall gain of speed by using the RISC processor.

Suppose the CPU is reading from a DRAM chip. The states of the eight transistors (reading from D7 to D0) are:

OFF   OFF   OFF   OFF   OFF   ON   ON   OFF

This is what one byte of a program really is. It is the *physical state* of a set of transistors or other electronic storage devices. When the byte is read, it becomes a set of high and low voltages on the data bus:

L L L L L H H L

To make it easier to read, we represent low voltage by '0' and high voltage by '1':

0 0 0 0 0 1 1 0

As the CPU works its way through memory reading the sequence of bytes, it finds:

00000110
01011001
01111000
00000110
00100011
10000000
01110111
01110110

This is what the CPU reads into its registers, one line (one byte) at a time. There, the corresponding bits in the register are either set (= 1) or reset (= 0). At this stage we are back to flip-flops that are set or not set. Remember, the program is really an array of set or reset fllip-flops in memory (or their equivalent, transistors that are switched on or off).

Depending on which bits are high or low in each byte, the CPU takes action accordingly. There are complicated logic circuits in the CPU which, on receiving a given combination of 0's and 1's, cause the CPU to perform a particular action.

Although they mean a lot to the CPU, rows of binary digits mean very little to us. The strings of 0's and 1's are difficult to read. To make them easier to understand, we take the rows of bits as binary numbers, and then convert them to their equivalents in hexadecimal:

06
59
78
06
23
80
77
76

This does not mean much, except to an expert in Z80 machine code.

Here is the list of codes with a explanation of what the CPU does as its reads them:

06	Load into register B the number that comes next.
59	$59.
78	Copy the value in B to the accumulator (A).
06	Load into register B the number that comes next.
23	$23.
80	Add register B to register A and leave the result in A.
77	Copy the contents of A into the address that is in the HL registers (we assume that there is already an address in HL).
76	Halt operations.

This table shows the contents of the registers and memory location at each step of the program.

Code	A	B	HL	$0173
06	0	0	0173	0
59	0	59	0173	0
78	59	59	0173	0
06	59	59	0173	0
23	59	23	0173	0
80	7C	23	0173	0
77	7C	23	0173	7C
76	7C	23	0173	7C

Notes: (1) numbers are in hex and the addition is hex.
(2) values are *copied* from one register to another, not moved.
(3) everything the CPU does is done in a series of many, short, simple steps.

The example shows how the CPU is told to add two numbers and store their sum in memory. Basically, all CPUs perform this operation in the same winding way. So all CPUs need a similar structure in order to do it.

Inside the CPU

Fig. 4.1 illustrates the main features found in most CPUs. The CPU comprises:

Control unit: This is a highly complex logic circuit which finds out what the microprocessor must do next and oversees the doing of it. There are several stages in its operation and it steps from stage to stage

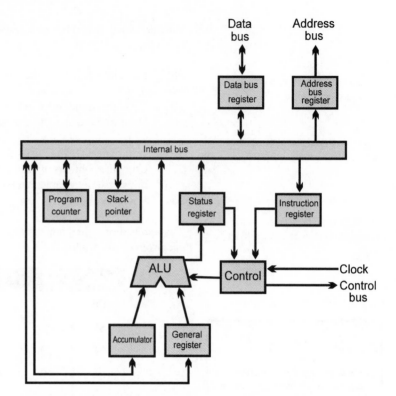

Register: an array of 8, sometimes 16 or 32, latches in which a byte (or 2 or 4 bytes) can be stored temporarily.

Figure 4.1 *The internal structure of a typical microprocessor includes complex logic circuits (control unit and ALU), with numerous registers, all linked by the internal bus.*

in response to pulses arriving from the system clock. It repeats this cycle indefinitely. Control lines, not shown in the figure, run from the control unit to all other parts of the microprocessor. Through these, it exerts its control over the whole MPU. It also has connections to external parts of the system (memory, and ports for example) through the control bus.

Internal bus: This has similar structure and performs similar functions to the address and data busses of the external microelectronic system. The internal bus may not have the same width as the address and data busses. For example, the Pentium has a 32-bit internal bus but has a 64-bit data bus. The 68000 family has a 32-bit internal bus, but the address bus has 24 bits and the data bus has 16 bits.

Data bus register: The register is the equivalent of an I/O port, through which the CPU receives data from the rest of the system or outputs data to it. Incoming data can be held in the register until the CPU is ready to receive it. Exchange of data between the internal bus and the data bus is under the control of the control unit.

Address bus register: It is the equivalent of an output port and used to transfer addresses from the internal bus to the external address bus.

Arithmetic/logic unit (ALU): The second complex logic circuit in the CPU is concerned with processing data. Its operations are controlled by the control unit. Given a single value it can increment it (add 1) or decrement it (subtract 1). Given two values it can add them, or subtract them. As explained in Chapter 7, multiplication and division is often done by repeated addition or subtraction. However, in some processors the ALU can perform multiplication and division directly, which is faster. The ALU is also able to perform logical operations, such as bitwise AND and OR, on a pair of values.

The remaining units of the CPU are registers. How many there are and the way they are used varies widely from one CPU to another, but there are four registers that are nearly always present:

Program counter: The CPU must always keep track of where it has got to in a program. The program counter is a register holding the most recently accessed address. As soon as the CPU has read the byte stored in that address, the address stored there is incremented by 1. We say that the program counter now *points to* the next address. This address is put on the internal bus, then into the address bus register and finally on to the address bus itself. In this way, the CPU works its way through a block of memory, reading each byte as it goes. Occasionally, the microprocessor has to jump to a different part of memory and continue its reading from a new address. On these occasions, the control unit puts the new address in the program counter, to direct the microprocessor to the different part of the program.

Stack pointer: Another register is used to hold the address of the top of the stack. The stack is a small block of memory where very important data is stored temporarily. The action of the stack is described in Chapter 11.

Status register: In some CPUs, this is called a *flag register*. It holds essential information about the result of an operation that has just occurred. The flag register usually has a capacity of one or two bytes, but the information is stored as the separate bits within the bytes. These *flag bits* are individually set to 1 or reset to 0 to indicate certain events. For example, every time a calculation gives a zero result, the *zero flag (Z)* is set to 1. If we want to know if the most recent calculation gave a zero result or not, the CPU can look at this particular bit in the status register to see if it is 1 or 0. Another flag, the *sign flag (S)* may be set when a calculation gives a positive result (zero counts as positive). A further example of a flag is the *carry flag (C)*, which is set when there is a carryout from a calculation. The use of this flag is illustrated in Chapter 7. The HC or *half carry flag* is set when there is a carryover

The addition in the program example above would be done by the ALU.

Bitwise: the corresponding bits in two values are operated on logically and the bit in the result is made 0 or 1 accordingly. There are examples in Chapter 7.

Pointer: When a register holds an address that is of particular importance, such as the next address to be read, we say that it is a *pointer* to that address, or that it *points to* that address. *Example:* the HL registers in the machine code program points to the address $0178, where the result of the addition is to be stored.

In the example on p.59, the program counter would hold 0 to begin with and then be incremented by 1 at each step until it reached 7.

Test your knowledge 4.2

What happens in the program on p. 59 if the sum of the two numbers is more than a single byte can hold?

Test your knowledge 4.3

What happens to the Z, S, and C flags after the sixth step in the program on p. 59?

BCD: binary coded decimal. A two-digit decimal number such as 47 is represented by 101111 in binary. In BCD the digits are coded separately in the high nybble and the low nybble, giving 01000111.

between B3 and B4. This is not used in ordinary addition, but is important if the program is adding in BCD. The table below shows the positions of these flags and others in the flag register F of the Z80 microprocessor:

Bit	7	6	5	4	3	2	1	0
Flag	S	Z	–	HC	–	P/V	N	C

Instruction register: When an instruction code is read from memory it is placed in the instruction register. The code is then passed on to the control unit which than carries out the instruction.

Accumulator and other registers: In many CPUs a special register, the accumulator (register A), is set aside as the main register used in processing. A high proportion of the instructions refers to operations involving the accumulator. We will refer to the accumulator as *A* from now on. Examples of operations involving A include loading data from memory into A, storing data from A into memory, incrementing and decrementing A, adding another stored value to one held in A, and manipulating data in A by shifting the bits in various ways. These things are done according to the instruction that has been stored in the instruction register. The result of the operation is placed on the internal bus, and is often circulated back to the accumulator to replace the value that was there before. At the same time, one or more of the flags in the status register are set or reset, depending on the result of the operation. Some operations involve two values, one stored in the accumulator and one in a general register. Fig 4.1 shows only one general register, but most microprocessors have several of these to make operations more flexible. The Z80 for example has two banks of eight registers. The main register set is:

A	F
B	C
D	E
H	L

A is the accumulator and F is the flag register. The remaining six registers can be used for storing data temporarily, for example the intermediate results in a sequence of calculations. All registers are 8 bits wide, but the six general registers can be combined in pairs BC, DE and HL to hold 16-bit values. The machine code program included an example of this for the HL pair where, as a 16-bit register, they were used to hold a RAM address.

There is an *alternate* set of 8 registers, called A', F', B' and so on, which can be switched over to when a second line of processing is required. The Z80 also has:

- Register I, which points to a table of interrupt service routines stored in memory.
- Register R, which counts the number of executed program steps.
- Registers IX and IY, which are two index registers used in indexed addressing (Chapter 12).

Fig 4.1 and the description above apply generally to many microprocessors and to many microcontrollers too. The main difference is that microcontrollers may also have memory and other devices such as timers built in to them. There are also CPUs, such as the 6502, which rely solely on the accumulator for processing. In other processors there is no special accumulator register set aside for the majority of the processing. Instead, the CPU has a number of registers, any of which can be subjected to the whole range of arithmetical and logical operations. For instance, the '1200' microcontroller has no accumulator but instead has 32 registers, any of which can be used in the same way as an accumulator.

Fast processors

On the whole, the modest processing speeds of microcontrollers are more than adequate for the tasks they have to perform. A washing machine that spends about 10 minutes at each washing stage and 5 minutes in rinsing and spinning does not need a processor capable of processing data at 10 million operations per second. In contrast, a flight control computer may have a mass of data to analyse as fast as possible before passing instructions to the control surfaces of the aeroplane. Similarly, a PC running a complex animated graphics program in over 16 million colours and with high-quality sound requires a high speed processor. These applications have led processor designers to increase the processing speed of microprocessors in various ways. An increase in the frequency of the system clock is an obvious solution, provided that the processor can be re-designed to operate at the increased speed.

Other measures to improve operating speed include:

RISC processors: These have been described earlier in the chapter. In most applications, they are faster than CISC processors. Some CPUs, such as the Pentium, can operate in either CISC or RISC mode depending on the requirements of the application.

Microcode

A CISC processor has several hundred instructions, some of them very complex. It is not possible for these instructions to be executed directly. Instead, when an instruction is waiting to be executed, the processor looks in a special ROM that is on its chip and finds there a short program that tells the processor how to carry out the instruction. In this way, the instruction is replaced by a *microprogram* of a special machine code known as *microcode.* The microcode takes over the operation of the control unit, the ALU, and other units of the processor until the microprogram is completed. Calling up, decoding and executing the microprogram takes time. The decoding may taken even longer that the actual execution.

RISC processors not only have fewer different instructions but all of these instructions can be executed directly as they reach the control unit. There is no microcode ROM on a RISC chip. This saves time, so nearly all instructions are executed in just one clock cycle. As a result, a RISC processor executes instructions about four times faster than a CISC processor.

Wide busses: Parallel transfer of data, both inside the CPU and in the system outside, allows more data to be transferred at each cycle. The faster processors have busses that are 32 or 64 bits wide.

Dual processing: Some processors have two ALUs working in parallel so that two instructions may be processed at once. This speeds up execution but this technique can not be used to full effect if the two instructions take different lengths of time to execute.

Prefetch buffer: In a conventional computer the instruction is fetched from memory, then executed. This is referred to as the *fetch-execute cycle*. The time required for fetching and executing (which may involve fetching data to work on), sets the speed of operation of the processor. A processor with a prefetch buffer (such as the 8086 family, including the Pentium) does not wait for the cycle to begin before fetching the instruction. Instead it takes any opportunity when it is not busy to fetch the next and subsequent instructions. It stores them in the buffer, without executing them. The instructions are then there, stored on the chip, ready to send on to the control unit as soon as the time comes to process them.

Cache memory: This is a special type of RAM with short access time. It may be located on the processor chip, so giving the fastest access, or there may be a special cache memory chip as part of the system's RAM. It is used for storing addresses and data that *might* be useful to the CPU in the near future. For example, it can hold data that has been read in ahead of the time it is required. When the CPU needs this data, it looks in the cache first. If it is there, it uses it. If it is not there, the CPU looks in the normal RAM. In some systems, data is stored in Level 1 (L1) and Level 2 (L2) caches. L1 is on the CPU chip so it is almost instantly available. The L2 cache is fast-acting DRAM on the computer board which is not so quickly read as L1 but is faster than the usual SRAM. Some processors have separate caches for data and for instructions. The instruction cache holds the data when it is first loaded and, from there, it goes to the prefetch buffer.

Floating point unit: The floating point format works with large positive and negative numbers that are stored in four bytes but is more complicated to process. Processing in the accumulator using software routines is slow. A floating-point unit is a logic circuit specially for processing numbers in floating-point format and is appreciably quicker. It may also include circuitry for multiplication and division.

Pixel: short for *picture element*. A picture on a screen consists of rows and columns of minute coloured dots. Typically a low-resolution screen is 640 pixels wide and 480 pixels deep, so the picture is made of just over 300 000 pixels.

MMX: *Multimedia extensions* are routines intended for speeding up graphics and sound processing in multimedia applications. They include SIMD instructions, which stands for *single instruction multiple data*. For example, a single SIMD instruction can be used to change the colour of many pixels at the same time.

Branch prediction: When a processor has performed a certain operation, it is more than likely that the program will require it to perform that same operation again. For example when the program has a loop in it, the processor has to jump back to the beginning of the loop every time it reaches the end of it. It may run round the loop hundreds of times. At any stage, it is safe to predict that it will jump back to the beginning of the loop when it comes to the end. It is only on the last time round the loop that this prediction will be wrong. Special routines within the CPU are used to store the instructions most recently used, on the assumption that they will be used again.

Loop: a portion of a program that is repeated several or many times. There are examples in Chapter 7.

Coordinated instruction set and compiler: As will be explained in Chapter 7, a compiler is a program that lets the programmer key in the program in a form that is more understandable than machine code. After it has been typed in, the compiler turns the program into machine code, that is, into instructions that the processor understands. Usually the instruction set is designed by the electronics engineeers who design the layout and logic of the chip. After that stage is complete, the compiler is written by a software designer. This may lead to problems when it is found difficult to write an efficient program for some of the instructions. For many of the more recent processors, including the '1200', the instruction set and the compiler are both designed at the same time. The hardware and software engineers work as a team. The result is a processor/compiler combination that operates with the best possible efficiency to give faster running programs.

Pipelining: In a conventional computer, a byte of data on its way from RAM to the ALU of the CPU is copied from one location to another (Fig. 4.2a). There is one transfer per clock cycle, so the whole process in the example takes 3 cycles. Bytes arrive at the ALU every 3 cycles. With pipelining (Fig. 4.2b) the second byte and third bytes begin their journey before the first byte has reached the ALU. A byte arrives at the ALU *every* cycle. Pipelining is advantageous but does not work well with certain kinds of instruction. Moreover, if the CPU is using branch prediction (see above) and the prediction is wrong, all the data in the pipeline has to be discarded.

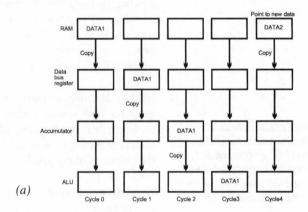

(a)

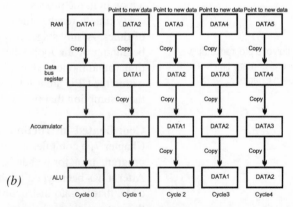

(b)

Figure 4.2 *Pipelining is a technique for speeding up the reading of data. Without pipelining (a), a byte may take 3 clock cycles to reach the ALU. A new byte arrives every 3 cycles. With pipelining (b), a new byte arrives every cycle.*

Out of order execution: If a calculation has several steps in it, the values to be used at a given step depend on values obtained in previous steps. The steps must be executed in the correct order. However, in a system with two or more processors, it is possible for an idle processor to check ahead to find an instruction that does *not* depend on previous calculations, and execute that instruction ahead of time. The result will then be ready when required.

Activity 4.1 Selecting a processor

Select a microprocessor or microcontroller that would be suitable for:

(a) switching traffic lights, with presettable delays at each stage.

(b) controlling an automatic weighing machine. Weights are displayed on a liquid crystal display.

(c) a hand-held stock logger, such as is used in supermarkets for shelf stock-taking.

(d) controlling a printer that is attached to a computer.

(e) monitoring the water level in a tank and warning when the rate of rise is too fast.

Use manufacturers' data sheets and other reference materials to help you make your choice. Make a list of the reasons for your choice.

Activity 4.2 Inside the CPU

Investigate the internal structure of a microprocessor or microcontroller, using the manufacturers' data sheets and other reference materials. Draw a diagram to show the main units and the way data flows between them. Show the ports and indicate the direction(s) of flow of signals. Write a brief account of the internal structure, outlining what each unit does.

List the flags of the status register (or equivalent register) and explain what each flag means.

Note any features of the device which suit it for special applications. Note any features that give the device high operating speed.

Problems on the CPU

1 What are the differences between CISC and RISC processors?

2 List the kinds of additional units that are included on the chip of several named microcontrollers.

3 What is machine code and in what forms does it exist in a microelectronic system?

4 Explain the function of the arithmetic logic unit.

5 Descibe the status register of the Z80 or other named CPU, and the meaning of any three of the flags.

6 List the registers of the Z80 or other named CPU, and describe their functions.

7 List four ways in which the speed of operation of a microelectronic system may be increased.

8 Write a general account of the architecture of a named microprocessor or microcontroller, illustrated by a block diagram. Explain how the architecture is related to the functions of the device.

9 Rewrite the table on p. 59 to show the computer performing the operation 172 + 35 (these are decimals) and storing it in $2A2C.

10 Using the information given in the example on p. 59, write a program in Z80 machine code to add 45, 125 and 23 and store the sum in $143D.

Multiple choice questions

1 The binary number 11010011101011 is represented in hex by:

      A $BE43.
      B $34EB.
      C $13547.
      D $D3A3.

2 A system is designed to have an address space from zero to $3FFF. The number of address lines required is:

      A 16 383.
      B 16 384.
      C 14.
      D 15.

3 A processor that is programmed by only a small number of instructions is called a:

      A RISC processor.
      B microcontroller.
      C CISC processor.
      D PLC unit.

4 Programs are stored in the SRAM memory of a computer as:

      A bits.
      B a series of 1's and 0's.
      C arrays of transistors switched on or off.
      D machine code.

5 The program counter of a processor holds $14C2. The processor reads an instruction that tells it to jump to a new address, ten bytes further on. The address in the program counter will change to:

        A $A.

        B $14C3.

        C $14CC.

        D $0000.

6 A register that holds the flags may be called a:

        A status register.

        B program counter.

        C index register.

        D instruction register.

7 If the zero flag is set (=1) it indicates that:

        A the result of the previous operation was not zero.

        B the result of the previous operation was zero.

        C the carry-out bit is 1.

        D the result of the previous operation as $0000.

8 With respect to CISC processors, RISC processors:

        A are slower.

        B may be faster in many applications.

        C are more difficult to program.

        D are faster.

9 A place where data is stored for fast access is:

        A cache memory.

        B DRAM.

        C the prefetch buffer.

        D a CD-ROM drive.

10 In bitwise logic, $59 AND $33 is:

        A $92.

        B $8C.

        C $7B.

        D $11.

5   Interfacing

Summary

Microcontrollers generally have built-in I/O terminals that can be directly connected to low-voltage, low-current peripherals. It is usually necessary to provide I/O ports to interface a microprocessor system to its peripherals. The ports may operate as parallel or serial interfaces. Ports may be built from TTL or CMOS logic or special ICs are available, such as the Z80 PIO and the Z80 SIO. Other I/O ports are discussed. To interface a digital system to an external analogue system there are analogue to digital converters for input to the digital system and digital to analogue converters for output. ADCs include flash converters and successive approximation converters. The DACs are represented by an R-2R converter.

A microelectronic system needs some way of communicating with the outside world. It needs *input*, by which it receives data and is told what to do. It needs *output* so that the processing of the data can have some useful result. Microelectronic systems communicate with the outside world through *ports*.

We have already seen that a microcontroller, such as the '1200' (Fig. 2.1) has two ports on its chip, known as Port B (8 bits) and Port D (7 bits). The bits of each port are individually programmable as inputs or outputs. Each port has a *data direction register* (DDR) where a series of 1s and 0s determine whether a bit is an input or an output. The DDR has an address in the memory space of the microcontroller. The program writes values into the DDR to set the pins as inputs or outputs. For example, if the value 11000000 is written into the DDR for Port B, the top two lines (B6 and B7) are set as outputs and the remaining lines as inputs.

As outputs, each line of the '1200' ports is able to sink 20 mA when the output is logic level low. When the output is high, it is able to source up to 4 mA. These figures assume that the IC is running on a 5 V supply. Provided the design keeps to these limits and that the total current sunk by the port is not more than 80 mA, it is safe to drive external circuits directly from the port. Some simple examples of this are shown in Figs. 2.6 and 2.7. In Fig. 2.7 we have not driven the filament lamp directly because such a lamp may require 60 mA or more. Instead, we have used a transistor as a switch, with resistor to limit the base current to a safe level.

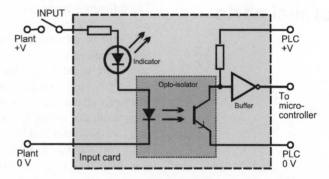

Figure 5.1 *A DC input interface of a PLC has an opto-isolator to allow the microcontroller to operate on an entirely separate DC supply, usually at lower voltage and free from noise. Here the sensor is a switch which might, for example, be a physical limit switch, a pressure switch, a tilt switch, or attached to a float. A PLC input card usually has 8 or 16 channels identical to the one shown here.*

As already mentioned, the sensors and actuators used in industrial applications often operate at voltages greater than that used for powering the microcontroller or may require current greater than a microcontroller output port can provide. It may be necessary to interface the microcontroller to AC circuits, and to circuits carrying an excessive amount of noise (see Chapter 6). PLCs have input and output interfaces that isolate the microcontroller. Fig. 5.1 is the circuit of a DC input interface. Similar precautions are taken in the circuits for DC output and for AC input and output.

A microprocessor does not have input and output ports. Instead, it connects with the other parts of the system through the data bus. However, the data pins of a microprocessor are not necessarily capable of sourcing or sinking sufficient current to drive the large number of devices that may be attached to the bus. Additionally, a system such as a microcomputer may be designed so that the CPU can drive the internal peripherals (keyboard and disk drives, for example), but could not be capable of driving external devices (printer and scanner, for example) that may be added to the system. Ports must be attached to the data bus to interface it to external devices.

Ports may be built from ICs belonging to the 74XX and CMOS families, or we may use special ICs intended to interface with particular CPUs. Ports are divided into two types:

- Parallel ports
- Serial ports

Parallel ports

The difference between parallel and serial transfer is explained in Figs. 2.2 and 2.3. Parallel ports are used where there is a large amount of data to be transferred and when the peripheral is not far from the CPU. For example, a printer is connected to the system by a parallel port. A serial port would be too slow for this application. Serial ports are used for long-distance communication, such as sending data by telephone. They may also be used where speed is not of prime importance and the complexities of having 8 parallel lines is best avoided. For example, a program is downloaded from a computer into a microcontroller using a serial port. Another example is a digital camera, where the image is fed to a computer through its serial port.

Parallel ports built from TTL

A port could be built from individual TTL gates but, more often, we use members of the TTL family specially designed for interfacing to microelectronic systems. One such IC is the 74244 octal *buffer* (Fig.5.2). To make the diagram simpler, we have drawn only one of the eight identical buffers. The function of the buffers is to make more power available for driving external devices. A typical TTL output of

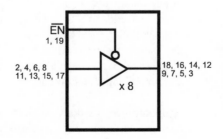

Figure 5.2 *The 74244 IC contains eight identical buffer gates with 3-state outputs. The outputs are enabled in two groups of fourgates when the two ENABLE inputs are made low.*

the 74LSXX series can sink up to 8 mA when the output is low. It can source up to 0.4 mA when the output is high. By comparison, for a buffer in the same series, the currents are 24 mA and 2.6 mA.

Note that the buffer shown in Fig. 5.2 is logic TRUE, or non-inverting. The output level is always the same as the input level. Inverting buffers are also available and may be useful in certain circumstances.

Buffers are used:

- as data output buffers. The inputs are connected to the data bus and the outputs to a data output socket or to the input of a peripheral device. The three-state outputs are normally not required in this application so the ENABLE pins are connected to the positive supply through a resistor. The peripheral device can read from the data bus at any time.

- as data input buffers. The inputs are connected to a data input socket and the outputs to the data bus. The outputs are enabled by a line from an address decoder only when the buffers are to place data on the bus.

- as address bus buffers. If there are many address decoders on the address bus, the CPU address outputs may not be able to drive them all. Instead, the CPU address outputs go to a buffer and from there the bus continues to the decoders.

Octal: Use to describe an IC that contains eight identical logic units.

Another type of port IC is an octal *latch*. An example is the 74LS373 illustrated in Fig. 5.3. This allows data to be stored (latched) at any time by making the ENABLE LATCH input low. In an input port, data from the peripheral can be latched at the instant the peripheral is ready to deliver it. The data is placed on the bus whenever the CPU is ready to receive it by making the ENABLE OUTPUT input low. Similarly, in an output port, data from the CPU can be transferred to a peripheral when the peripheral is ready.

A third type of port uses an octal D-type *flip-flop*, such as the 74LS374. The pinout of the IC is the same as that of the 74LS373 (Fig 5.3) except that the EL input is replaced by a clock input.

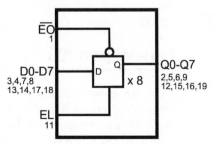

Figure 5.3 *There are eight identical latches in the 74373, each with a data input (D) and a 3-state output (Q). The latches are all controlled by the ENABLE LATCH and ENABLE OUTPUT inputs.*

Latches and flip-flops

Both of these are data storage devices but they act in slightly different ways:

Latches: There may be several (often 8) latches on the same chip and these are all controlled by the $\overline{\text{STORE}}$ or $\overline{\text{LATCH ENABLE}}$ input. While the $\overline{\text{STORE}}$ input is high, data at the output of each latch follows the data at the input. When the $\overline{\text{STORE}}$ input is made low, data present at the output at that instant remains unchanged. It remains unchanged until STORE goes high again, when it then follows input again.

Some types of latch have three-state outputs. Some latch when the STORE input goes high, instead of low.

Flip-flops: The type used in ports is the D-type and there may be several (often 8) on the same chip. They are controlled by the CLOCK input. Data at the output of the flip-flop remains unchanged until the CLOCK input goes from low to high. The data present at the input at that instant is then transferred to the output. There is no stage at which output follows input, as in the latch.

Summing up:

Latches follow or latch.
D flip-flops change only at the clock edge.

Test your knowledge 5.3

What is meant by the 'XX' in '74LSXX'?

We may also build a port using an octal bus *transceiver*. As the name implies, they provide two-way buffered communication between two busses. One may be the internal data bus of a computer; the other may be the bus of a microelectronic system interfaced to the computer. Control inputs set the device to operate in one direction or the other.

The 74LSXX ICs mentioned in this section are also available in CMOS versions.

Parallel port ICs

There are several different types of parallel port IC with very similar features and a few minor differences. Fig. 5.4 shows an IC intended for use with microprocessors of the Z80 family, the Z8420, more usually known as the Z80 PIO (parallel I/O).

Fig 5.4 illustrates the more important features of the Z80 PIO. It has a bi-directional connnection with the system data bus which is 8 bits wide. It has two 8-bit ports, referred to as Port A and Port B. Each port has two handshaking lines for communicating with devices attached to the ports. The lines are RDY (ready) and $\overline{STB}$ (strobe).

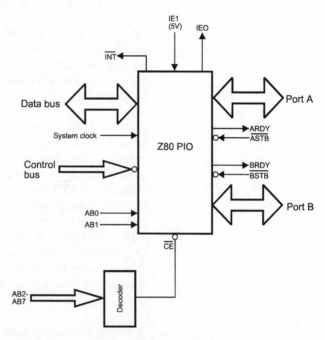

Test your knowledge 5.4

Write out the address $F2 in binary.

Fig. 5.4 *The Z80 PIO provides two ports which may be configured as inputs or outputs.The decoder enables the IC when the upper six bits of an address in the range $F0 to $F3 is presented to it. Inputs marked with small circles are active-low.*

The PIO operates in one of four modes for each of the ports:

Mode 0   As an 8-bit output port
Mode 1   As an 8-bit input port
Mode 2   As an 8-bit bidirectional port (Port A only)
Mode 3   Port B as a bitwise I/O port when Port A is in
            mode 2.

Each port has a port control register to which a byte is sent to select the mode. If the port is programmed as mode 3, a second byte must be sent to define which bits are inputs and which are outputs.

The port registers and port control registers are allocated a range of four addresses. These are fixed addresses which are automatically put on the bus by the Z80 CPU when accessing the port. They can not be chosen by the user. The addresses are:

$F0   Port A data register.
$F1   Port A control register.
$F2   Port B data register.
$F3   Port B control register.

All four addresses have the same top six bits (111100). The decoder circuit receives address lines AB2 to AB7 and makes the $\overline{\text{CE}}$ input low when these carry 111100. The register selected depends upon the bottom two bits, AB0 and AB1 which go directly to another decoder inside the PIO.

In Mode 0 (the port as a data output), the CPU writes data into the data register. This causes the RDY output to go high, which tells the peripheral that there is data in the register waiting to be read. The peripheral reads this data when it is ready to do so, and then puts a low pulse on the $\overline{\text{STR}}$ line. This tells the PIO that the data has been read and this message must now be passed on to the CPU. The PIO first makes the RDY output low, as there is no new data to read. Then the $\overline{\text{INT}}$ output of the PIO is made low. This interrupts the CPU, which is programmed to respond in some way, usually to send the next byte of data. By using this simple handshaking procedure, successive bytes of data are transferred from the CPU to the PIO and then to the peripheral.

A similar technique is used in Mode 1 (the port as a data input). If the PIO is ready to receive a data byte from the peripheral, its RDY output is high. The peripheral puts data on the port bus, then sends a low pulse on the $\overline{\text{STR}}$ line. This latches the data in the port register. The RDY line is made low so that the peripheral does not send another byte until this byte has been read by the CPU. The $\overline{\text{STR}}$ pulse causes the PIO to interrupt the CPU by putting a low level on the $\overline{\text{INT}}$ line. The CPU is programmed to read the data from the register when it is interrupted. The control signals at the end of the read operation cause the RDY line to go high, indicating to the peripheral that it can now send the next byte.

The above description shows how the data is transferred with handshaking, so that (if inputting) the CPU will not miss any data. Conversely, when outputting, the CPU will not send any data until the previous byte has been acknowledged. These procedures can be simplified in some circumstances. At an input port, the $\overline{\text{STR}}$ line can be held permanently low (perhaps connected directly to the 0 V line). Then any data that the peripheral sends appears immediately in the register and can be read by the CPU at any time. For example, the

Other parallel port ICs

There is a wide range of parallel port ICs available, similar to the Z80 PIO, but usually specialised to work best when connected to one particular CPU.

The 8255A PPI (programmable parallel interface) has three 8-bit ports programmable as inputs or outputs. The individual pins can not be programmed, but the high and low nybbles of Port C can be programmed separately. Handshaking is provided. The PPI has the same addressing as the Z80 PIO, except that addresses $F0 to $F2 cater for the three ports and all control signals go to $F3. As in the Z80 PIO, all mode control operations involve sending a byte to the control register. The bit set/reset mode is of interest. It allows the bits of Port C to be set or reset. Only one bit can be set or reset at one time. This feature is useful for producing strobe signals.

The Intel 68230 PIT (parallel interface timer) is designed for 68000 systems. Port A and Port B are both definable as 8-bit input, output or bidirectional ports. Alternatively, they may be combined as a single 16-bit input, output or bidirectional port. The ports may also be programmed for bitwise operation. There is a set of lines for handshaking. There is a third port, Port C, which can be used as a I/O port but many of its pins are multipexed for other functions, including acting as input and output for the on-chip timer.

The timer is a 24-bit counter that has a value loaded into it and then counts down. Depending on the programming, it triggers various events when the count reaches zero. The timer can be programmed to generate periodic interrupts, a square wave of selected frequency, or a single interrupt afer a preset period of time. In addition to these functions, it can be used to measure elapsed time. As is usual with such devices, the timer is programmed by the CPU writing codes into its registers. The registers can also be read to discover, for example, the length of time elapsed since it was reset.

peripheral could be a DAC, continuously providing data, which is sampled by the CPU at regular intervals. The output routine may similarly be simplified. Data sent by the CPU is always immediately latched into the register and can be read by the peripheral at any time.

The peripheral reads the data whenever it needs it, ignoring the RDY line and not signalling back on the $\overline{\text{STR}}$ line. The $\overline{\text{INT}}$ routine is disabled in the CPU so that it sends data whenever it has data to send.

In Mode 2, Port A is bidirectional, and four handshaking lines are available. The operating method is a combination of the input and output sequences described above. When Port A is in mode 2, Port B must be assigned to Mode 3, the bitwise I/O. There is no handshaking on Port B in this mode. Signals placed on the output pins by the CPU go straight out to the peripheral. Signal placed on the input pins by the peripheral are put on to the data bus when the PIO is enabled. In Mode 3 it is possible to configure all eight lines in the same direction, in which case the port becomes an input or output port with no handshaking. Alternatively, the pins may be configured separately and up to eight 1-bit inputs or outputs may be connected to the port.

Serial ports

For transmission of data over short distances of a metre or two, an ordinary TTL or CMOS gate or buffer can be used. Additional gates are needed if any handshaking signals are to be transmitted as well.

For longer distances it is preferable to use a serial port based on the RS-232 standard. This allows for transmission over distances up to 15 m. The driver output signal is between ±5 V and ±15 V. The maximum data rate is 20 Kbit per second. Other RS standards exist for higher rates of transmission. The RS standards specify the types of connector to be used and the functions of each signal line.

The standard specifies the use of nine lines for carrying the serial signal and the handshaking signals. Fig. 5.5 shows the connections between a transmitting computer and receiving device, such as a modem. A 25-pin D-type connector is used at both ends of the cable. More recently, a 9-pin D-type connector has been used. The pin connections are:

Signal	Signal name	9-pin	25-pin
TD	Transmitted data	3	2
RD	Received data	2	3
RTS	Request to send	7	4
CTS	Clear to send	8	5
DSR	Data set ready	6	6
SG	Signal ground	5	7
CD	Carrier detect	1	8
DTR	Data terminal ready	4	20
RI	Ring indicator	9	22

When using a 25-pin socket there is often a line called Protective Ground connected to the metal chassis of the terminal at either end. This uses pin 1. TD is sometimes known as TXD and RD as RXD. SG is sometimes called GND.

In practice, many systems use the 9-pin connectors with only three lines, TD, RD and SG for two-way communication. One-way connection needs only two lines; TD at the transmitting end is wired to RD at the receiving end.

Interfaces can be built from TTL gates, as in Figs 5.5 and 5.6. The TTL input to the interface (Fig. 5.5) comes from the transmitting system, possibly through its output port. The voltage levels of the system are 0 V (= logic 0) and +5V (= logic 1). The 74LS06 gate inverts these levels. Note that the gate has an open collector output, requiring a pull-up resistor. At this stage, the voltage for logic 0 is *higher* than that for logic 1. As the signal passes through the circuit it is inverted by each of the transistors so is still inverted at the output. Here the levels are ±12 V, with +12V being equivalent to logic 0 and −12 V being equivalent to logic 0. These levels are valid RS232 levels.

Open collector output: The output transistor of the gate has no collector resistor, so one must be provided externally. The resistor may be connected to any positive voltage (maximum 30 V). This gives a high output voltage that is higher than the standard +5 V.

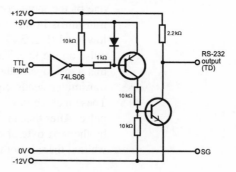

Figure 5.5 *This interface converts a TTL signal (0 V = 0, +5 V = 1) to an RS232 signal (−12 V = 1, +12 V = 0).*

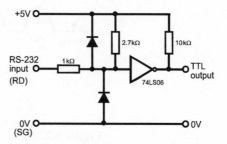

Figure 5.6 *The INVERT gate used in this RS-232/TTL interface has an open-collector output.*

At the receiver, the signal first passes through a network comprising a resistor and two diodes to limit the voltage swing to between 0 V and +5 V. This is fed to the 74LS06 gate which inverts it, so restoring the original logic levels. Now +5V again corresponds to logic 1, and 0 V to logic 0. Once again, the gate needs a pull-up resistor. Its TTL signal may now be fed to any TTL input, including the input port of the receiving system.

A number of ICs are available for transmitting and receiving RS-232 signals. An example is the MAX232 IC, which provides for two RS-232 lines in each direction. In includes a voltage-doubling circuit to generate +10 V from the +5 V supply. This means that only the usual +5V supply is needed to produce the RS-232 levels in the chip. It also includes an inverter circuit to generate −10 V. The IC has two TTL to RS232 inverter gates and two RS232 to TTL inverter gates.

Protocol: an agreed set of rules.

When data is being sent without using handshaking lines, it is important to have a protocol for data transmission. RS-232 does not include any such protocol but there are a number of data transmission protocols agreed by international bodies. These are used not only for RS-232 transmissions but for other interfacing techniques as well. In one of the more common procedures, a transmitter that is waiting to transmit is in the 'marking' state. Its output signal is at a continuous low level (Fig. 5.7). As soon as it has data to transmit, it places a high pulse on the line. This is called the Start pulse and warns the receiver that a message is about to arrive. Immediately after the Start pulse, the transmitter sends eight data bits, of length equal to the start pulse. These may be low level or high level, and are followed by a low Stop pulse. After this, the transmitter may send another Start pulse followed by the next byte of data, or it may remain in the 'marking' state for a while if there is nothing to send.

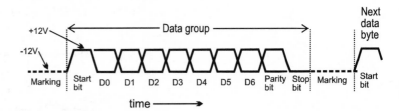

Figure 5.7 *This serial interface protocol is called* asynchronous *because it does not need the transmitter and receiver to have their system clocks synchronised. In some systems the Stop bit must be at least two bit-periods long.*

Bit rate and baud rate

Both of these are used to express the rate of transmission of data. They are not the same thing.

Bit rate is the number of binary bits (0's or 1's) transmitted in 1 second. Faster tranmission rates are often expressed as kilobits (1024) bits per second.

Baud rate is the number of 'signal events' transmitted per second. If one signal event (for example, a pulse of given length or amplitude) represents one binary digit, the baud rate and the bit rate are equal. But in many systems a signal event represents more than one bit.

The eight data bits usually consist of seven data bits, D0 to D6 followed by a *parity* bit (see box). For data to be read correctly, the transmitter and receiver must both be operating at the same baud rate. However, there is no need for precise timing because each group begins with a Start pulse, which resets the clock of the receiver. It then has to keep time for only 8 pulses. This method of signalling is described as *asynchronous*, because it does not require the clocks at the transmitter and receiver to be permanently synchronised.

Right shift: The direction 'right' applies to data in which the LSB is to the right and the MSB to the left.

Data is processed in parallel form in microelectronic systems. Before it can be sent to a serial interface such as the one shown in Fig. 5.5 it must be converted to serial form. There are several ICs that can do this, including the 74LS165 parallel-serial converter. This IC is described as a *shift register* (Fig. 5.8). It has 8 registers, A to H, each of which holds one bit of data. The registers are connected internally so that the data in one register can be shifted into the register on its right. It has 8 data inputs which, in a computer, could be connected to the data bus.

Test your knowledge 5.5

Data bit D5 is loaded into the shift register and shifted right three times. What register is it then in?

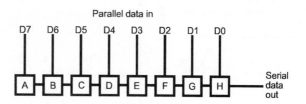

Figure 5.8 *A shift register is used to convert parallel data into serial data, ready for transmission by a serial output interface. This is a parallel in serial out (PISO) shift register.*

When its $\overline{\text{LOAD}}$ input is made low, the data present on each data line is loaded into the corresponding register. The IC has a data output and an $\overline{\text{ENABLE}}$ input. If the $\overline{\text{ENABLE}}$ input is held low, the data is shifted one step to the right every time the clock input rises from low to high. On the first rising clock edge, the data in H (D0, equal to 0 or 1) appears at the serial output terminal and can be sent to a serial output interface. The data in G is transferred to H, and the other transfers are F to G, E to F, D to E, C to D, B to C, and A to B. Each bit has been shifted one place to the right. On the next rising clock edge, D1, originally in G but now in H, appears at the serial output and goes to the output interface. The other data bits are shifted one step to the right. The process repeats for the next six clock edges, the bits appearing at the serial output in order D2, D3, D4, D5, D6, and D7.

Parity

This is a technique for detecting errors in transmission. When a group of 7 data bits is transmitted the number of 1's is counted and an extra bit, the parity bit is added at the end, so that the total number of 1's is even.

Examples:

> If the seven bits are 1101010, the group already contains an even number of bits. So the parity bit is 0 and the group transmitted is 11010100.

> If the seven bits are 0110111, the group contains an odd number of bits. The parity bit is 1, to make the number of bits even. The group transmitted is 01101111.

The receiving system counts the number of bits in each group and rejects any group that contains an odd number. Then the parity bit is removed from the end of the group and the remaining seven bits are sent on for processing. The receiver may be able to request the transmitter to send any rejected groups again.

This technique is known as *even parity*. Some systems use odd parity, in which the number of 1's is made up to an odd number.

Parity checking will detect a single error in a group, but does not detect two compensating errors. Adding a parity bit to each group means that the rate of data transfer is slightly reduced.

Test your knowledge 5.6

If parity is even, what is the parity bit (X) for these 7-bit groups?
(a) 1101100X.
(b) 0111010X.
(c) 1110000X.

More on parity

Other more complicated systems of parity checking have been devised. In the system outlined below, 16 bits of data are arranged in four groups of four, and a parity bit is added to each row and column. Five groups of five bits are transmitted instead of four groups of four. This means that the rate of data tranfer is reduced, but there are advantages.

Example :

This example is worked with even parity.

	Data				P
D	1	1	0	1	1
a	0	1	1	1	1
t	1	0	0	1	1
a	0	1	1	0	0
Parity	0	0	0	1	1

In the table above, it can be seen that the parity is wrong for the second column and the third row. This locates the incorrect bit, which is in the shaded box. The analysis locates the incorrect bit, which can then be changed automatically, to a 1 in this case. There is no need to repeat the transmission.

This technique also checks that the parity bits have been received correctly:

	Data				P
D	1	0	0	0	1
a	1	1	1	0	0
t	0	0	0	1	1
a	1	0	1	0	0
Parity	1	1	0	1	1

Parity is wrong in the 2nd row and 5th (parity) column.

At the receiving end of the transmission, the serial data must be re-formatted into parallel data so that it can be processed further. Here we need a serial-to-parallel converter, such as the 74LS164 (Fig. 5.9). This has a serial input and eight data outputs. The way it works is the converse of the previous shift register. Data is clocked in starting with D0 and ending with D7. D0 is first stored in register A. As clocking proceeds D0 is right-shifted along the registers until it reaches register H, with all the other bits in order along the chain. Then the whole byte is unloaded through the eight parallel outputs. The data can be fed directly to a device such as a decoder that controls a 7-segment LED display. If it is to go on to a data bus, it must first be stored in a set of latches with three-state outputs.

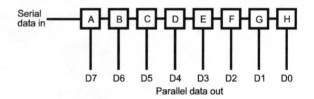

Figure 5.9 *Received serial is converted back to parallel form by a serial in parallel out (SIPO) shift register.*

Shift registers perform the essential conversion between serial and parallel data but in most systems it is necessary to add start, stop and parity bits before transmission. On reception, it is necessary to remove the start and stop bits, to check parity and finally remove the parity bit. Although parity checking ICs are available, the logic circuit required for this degree of processing is very complicated and it is easier to use a ready-made IC. Most CPU families include a serial output interface IC that accepts parallel data and produces serial data complete with the additional bits. They also accept serial data, check its parity and produce a parallel output.

The Z80 SIO (serial input/output) is an IC that includes all the necessary facilities on the same chip (Fig. 5.10). The SIO is a very versatile IC and has several operating modes. We will look at the features that are of general interest. The SIO has two channels, A and B that are separately programmable. As with the PIO, the CPU communicates with the IC as if it consists of four registers, a data register and control register for Channel A, and the same for Channel B. The CPU controls the SIO by writing bytes into the control registers.

With serial communications it is essential that both the transmitting

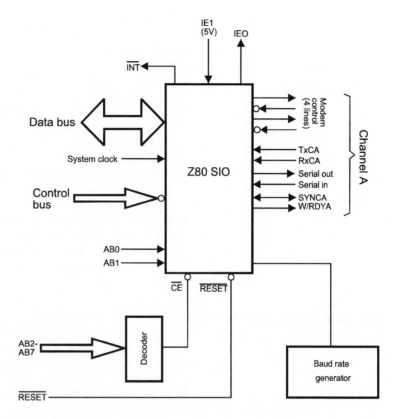

Figure 5.10 *Comparison with Fig. 5.4 shows that many of the connections of the Z80 serial input/output device are the same as for the Z80 PIO. In the right the figure shows only Channel A.. The lines of Channel B are identical except that it has no TxC and RxC inputs. Both channels share the baud rate generator.*

station and the receiving station are set to receive data at the same rate. This is usually specified by stating the baud rate (see box, p. 81). There are a number of standard baud rates, ranging from 110 to 38400. The baud rate generator may be a simple crystal-controlled TTL oscillator or there is the Z80 CTC generator. Also available are special baud rate generators such as the MC14411. The timing chain of the SIO can be set to divide the pulse rate by different amounts so that it is possible to send and receive at a selection of baud rates without having to change the clock.

The SIO can be programmed (by coded bytes written into its control registers) to operate according to a number of different protocols (see Fig. 5.7). It is possible to select for a character of 5, 6, 7, or 8 bits, and to add 1, 1½, or 2 stop bits. It is also possible to choose between even parity, and to disable the addition of a parity bit if preferred.

If the system is to communicate with another system by way of the public telephone network, it is necessary to use a *modem*. This is a device which receives a serial transmission of high and low logical levels (as in Fig. 5.7) from the SIO and converts it into a form suitable for transmission over the telephone line. This is known as *modulation*. The demodulator section of the equipment converts the received modulated signal back into pulses at high and low logic levels. One modulation technique is *frequency shift keying* (FSK). The logic lows and highs of the serial data are represented in the modulated signal by two different audio frequencies. For example, in the Kansas City protocol, a 0 is represented by four cycles at 1200 Hz, and a 1 is represented by eight cycles at 2400 Hz. Thus, each bit takes the same length of time to transmit. Control signals between the SIO and the modem of each channel are sent by way of four handshaking lines. There are also two lines for sending and receiving the serial data.

Modem: short for
MOulator-DEModulator.

Data converters

Some systems need to be able to accept analogue input. The input (usually a voltage) varies smoothly over a given range. The output must be a digital quantity. For this purpose, the input interface includes an IC known as an *analogue-to-digital converter* or *ADC*. Examples of systems needing an ADC for input are audio systems (such as a digital tape recorder) and many instrumentation systems (such as a digital multimeter).

There are several types of ADC, but the two most popular types are flash converters and successive approximation converters.

Flash converters

Flash converters are faster than the other types but have the disadvantage of being more expensive than other converters with comparable precision. Fig. 5.11 demonstrates why. A flash converter consists of a number of comparators connected by their inverting (–) inputs to a chain of resistors. The chain of resistors is connected at one end to 0 V and at the other end to a reference voltage. This produces a fixed range of voltages along the chain. The input voltage is fed to the non-inverting (+) inputs of all the comparators. Each comparator compares this voltage with the voltage it is receiving from the chain. The outputs of the comparators then swing either low (0) or high (1), depending on whether the input voltage is less than or greater than the voltage from the chain. The result of this is that, as the input voltage increases from 0 V, the outputs become 0000000, 0000001, 0000011, 0000111, and so on, to 1111111. There are 8 possible outputs, which are sent to a

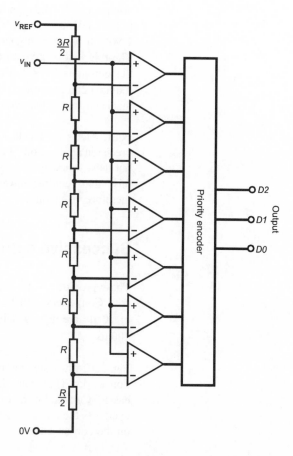

Figure 5.11 *A flash analogue to digital converter is based on a chain of resistors connected to an array of comparators.*

priority encoder. This is a logic circuit which determines which is the highest '1' bit (counting from the LSB on the left, that is, from the bottom of the chain). The output of the encoder ranges through all the binary values from 000, 001, 010, 011, and so on to 111. This output is proportional to the input voltage, and the conversion of analogue to digital is complete.

The conversion time of a flash converter is the time taken for the comparators to settle, plus the propagation delay in the gates of the encoder. Typical conversion times lie in the range 10 ns to 2 μs, which is fast enough for the conversion of audio signals in real time.

The ADC of Fig. 5.11 has seven comparators and there are only 8 possible output readings. This has to cover the entire input range from 0 V up to the reference voltage. As a simple example, if the reference voltage is 8 V, the eight possible values from the encoder correspond to 0 V, 1 V, 2 V, ... , 8V. We can read the voltage only to the nearest 1 volt. If we want higher precision, we must have more comparators. The rule is that, given n comparators, the number of possible steps in the output is $2^n - 1$. For example, a flash ADC with 8-bit output requires 256 converters. Such an ADC often has a 2.56 V reference, so that each step in the output is 0.01 V.

Flash ADCs are made with 4-bit outputs and 8-bit outputs for low-precision applications. For higher precision there are 12-bit converters, but these work by a technique known as *half-flash*. This is a compromise that requires fewer comparators. It works in two stages and therefore takes longer.

Successive approximation converters

These provide greater precision than flash converters at relatively low cost. Converters with 16-bit precision are available. Conversion times of 20 μs are achieved by some types, though some take as long as 100 μs.

Fig. 5.12 illustrates the principle on which the successive approximation works. This example has only four bits, to make the explanation easier. Conversion takes place in a number of steps, one per clock cycle. As the name suggests, the converter works by gradually homing on the correct output. To begin, the START CONVERSION input of

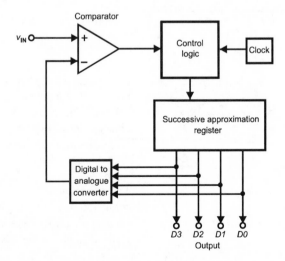

Figure 5.12 *At each stage of conversion, the analogue equivalent of the output value in the SAR is compared with the analogue input voltage.*

the IC is made low. At the next clock step, the control logic sets the first bit (the MSB) of successive approximation register (SAR) to '1'. In other words, the first approximation (or guess) at the correct output is '1000', the 'half-way point'.

There is a digital-to-analogue converter that converts this approximation to the equivalent analogue voltage. This is then compared with the input voltage. If the guess is less than the input voltage, we know the approximation is too low. The MSB stays at '1' and the control logic goes on to try the next digit. If the guess is more than the input voltage, it is too high. The MSB is reset to zero, and the logic goes on to try the next digit.

If the guess is too low, setting the next digit gives 1100 from the SAR. This too is converted and tested against the input voltage. Again, the digit is either retained as '1 'or reset to '0'. The routine is repeated at each clock cycle, working along from the MSB to the LSB. After 4 cycles, conversion is complete and the END OF CONVERSION output of the IC goes high. The data is then present on the parallel outputs of the converter. With some ADCs, the data is presented at a serial output, MSD first. The time taken for the conversion is roughly proportional to the number of bits in the output, so higher precision is obtained at the cost of longer conversion time.

One of the problems with successive approximation is that a rapidly changing input may be impossible to evaluate. The homing routine does not work properly with a moving target. If this is a problem, a sample and hold circuit is used to sample the input voltage and hold it while it is being converted.

Digital-to-analogue converter

A system that has processed data in digital form may need to output it in analogue form. For example, the circuit of a CD player has to output an audio signal, which is an analogue quantity. This requires a *digital-to-analogue converter*, or *DAC*. There are several types of DAC but the commonest is based on the *R–2R ladder*. Fig. 5.13 shows a 4-bit example. The switches are actually CMOS switches under logical control.

The most frequently used DACs depend on a resistor network usually known as an R-2R ladder. Typically, R equals 10 kΩ. There are CMOS switches S0 to S3 in each 'rung' of the ladder which can be switched either to the 0 V line or to the inverting (−) input of the operational amplifier. Whichever way a switch is set, the same current flows out from the rung, so the currents flowing in the network are not affected by which switches are open and which are closed. A short calculation

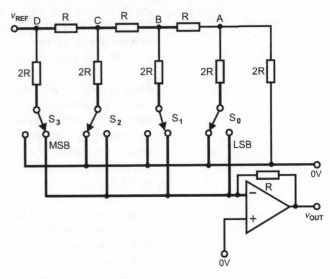

Figure 5.13 *The R-2R 'ladder' produces currents that are weighted on the binary scale. They are summed by the operational amplifier.*

In a binary number 1111, the MSB represents 8 decimal, and the following bits represent 4, 2, and 1 in that order.

If the DAC described here had an input of 01100101, what would be its output voltage?

shows that the current flowing out of a 'rung' is equal to half of the current flowing out of the 'rung' to its left. For example, if the reference voltage is 5 V, and 2R is 20 kΩ, the current flowing through S3 (the MSB switch) is 250 μA. The current flowing through the other 3 switches is 125 μA, 62.5 μA and 31.25 μA. The currents through switches S3 to S0 are proportional to the values of the corresponding bits D3 to D0 in a 4-bit binary number. Similarly, the sum of the currents is proportional to the sum of the bits. In the figure, S3 and S1 are switched to the op amp, so the equivalent binary number is 1010, and the sum of the currents is 250 + 62.5 = 312.5 μA.

The op amp is connected as an inverting summer. Its output swings negative by an amount proportional to the sum of the currents. It takes only a second inverting amplifier with a fixed gain of −1 to make the output positive. The output appears as a voltage, proportional to the reference voltage.

Precision

The precision of the flash and R-2R converters depends on the precision of the resistors. Their exact values do not matter but it is important for the *ratios* between them to be exact. This is relatively easy to achieve because all the resistors are fabricated at the same time on the same chip.

R-2R DACs convert in the short time it takes for the logic to set the switches and for the op amp to settle. DACs are available for inputs up to 16 bits wide, and a settling time often less than 1 μs.

Activity 5.1 ADCs

Study the data sheet of an analogue to digital converter and note its main operating conditions. Set up the ADC on a breadboard and provide it with:

- A suitable power supply.

- A variable input voltage for conversion.

- A set of 8 LEDs switched by transistors to display the digital output.

- If it is a successive approximation converter, it will need a clock pulse generator, a low pulse generator at the START input (use a 10 kΩ resistor to pull the pin up to +5 V, and connect a push-button between the pin and 0 V). Use an LED indicator on the EOC output.

The data sheet may describe other connections that should be made.

Make conversions at a number of input voltages between 0 V and the supply voltage. Plot a graph of the digital output reading against the analogue input voltage. Comment on the curve you obtain.

Activity 5.2 DACs

Study the data sheet of a digital to analogue converter and note its main operating conditions. Set up the DAC on a breadboard and provide it with:

- A suitable power supply.

- A set of 8 switches to provide each data input with 0 V or the positive supply voltage.

> • A multimeter to measure the output voltage.
>
> The data sheet may describe other connections that should be made.
>
> Make conversions at a number of digital inputs between 0 and 255 (assuming an 8-bit input). Plot a graph of the analogue output reading against the digital input. Comment on the curve you obtain.

Interfacing to a PC

When we interface to a computer, we are not connecting directly to the CPU as described in other parts of this chapter. The computer is a complete microelectronic system and it has ports to which we can connect external circuits of our own design. The PC usually has one parallel port and one serial port, though some models may have more. The most appropriate port for the interfacing projects described in this book is the parallel port, usually known as LPT1. This is the port to which a printer is normally attached. It may have other devices daisy-chained to it, such as an external disk drive.

The standard parallel port has a D-type 25-pin socket on the rear of the computer. The pins of this socket are allocated to three ports, as shown in Fig. 5.14. There are eight ground lines, of which one must always be connected to the 0 V line of any device attached to the computer.

As might be expected, the ports are allocated addresses in the PC's memory space. To output and input data we send it to the address of one of the ports. Using a high-level language, we do not have to deal with data direction registers or with handshaking. All this is taken care of by the machine code generated by the running of the program. In Chapter 10 there are some programs which demonstrate how this is done using BASIC.

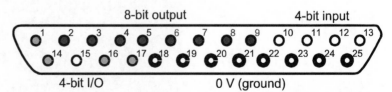

Figure 5.14 *The parallel port of a PC is seen as a 25-pin socket. It has pins for three ports and eight ground lines. In the figure the socket is viewed from the rear, as you see it when you are about to plug in the printer cable.*

This table shows the addresses used for sending or receiving data from the three registers:

8-bit output	4-bit input	4-bit I/O
$03BC	$03BD	$03BE
$0378	$0379	$037A
$0278	$0279	$037A

The correct address to use can be found by looking in System Information.

Referring to Fig. 5.14, the pins of the 8-bit register are as follows:

Pin	9	8	7	6	5	4	3	2
Bit	D7	D6	D5	D4	D3	D2	D1	D0

After data has been loaded into this register, it can be read back to check it.

The 4-bit output register is more complicated. It is an 8-bit register but the lower 4 bits are not used for data output:

Pin	11	10	12	13	–	–	–	–
Bit	D7	D6	D5	D4	D3	D2	D1	D0

Bit 7 is inverted. If its input line is low, D7 equals 1. This bit causes an interrupt when it is made high (that is, its input line is made low). However, it does not do this if it is disabled by making bit 4 of the 4-bit I/O register low.

The 4-bit I/O register has the following pin allocations:

Pin	–	–	–	IE	17	16	14	1
Bit	D7	D6	D5	D4	D3	D2	D1	D0

Bits 0, 1, and 3 are inverted. Bit 2 is normal. To use this register as input, all bits D0 to D3 must first be made high. Then input data either pulls the bit register low, or leaves it high. It is then read.

Apart from the complications mentioned above, the ports are suitable for the 1-bit interfacing described in Chaper 10. For more extensive interfacing there are I/O cards that can be plugged into slots inside the computer, and which have their own special addresses and decoding circuits.

Problems on interfacing

1 Describe and compare two different output interfaces.

2 Describe and compare two different input interfaces.

3 List and describe the characteristics of three TTL devices that can be used as input/output ports.

4 Explain how you would use a Z80 parallel input/output IC (or any named device that is similar) to interface a microprocessor to (a) a printer and (b) a 4-key keypad.

5 What hardware would you use to send signals from a CPU over the telephone network?

6 What is an analogue to digital converter? How would you interface it to a named processor? Give an example of a practical use for such a circuit.

7 Describe how you would interface a digital to analogue converter to a named processor.

8 Explain the difference between a latch and a flip-flop.

9 Describe the action of a shift register. For what purposes are shift registers used in microelectronic systems?

10 Explain the difference between bit rate and baud rate. When are they equal?

11 What is meant by parity? How and why is it checked in serial data transmission?

12 Describe a method for interfacing devices to a personal computer. How could you use this interface to receive data from an ADC?

Multiple choice questions

1 When it has a logic low output, a TTL gate can sink up to:

    A 24 µA.
    B 8 mA.
    C 1 mA.
    D 16 µA.

2 Buffers are used is certain systems because they:

    A prevent voltage spikes from passing.
    B change state very rapidly.
    C change a logic high to a logic low.
    D sink and source larger currents.

3 The RS-232 standard covers serial transmission for distances up to:

    A 15 m.
    B 1 km.
    C 200 m.
    D 8 m.

4 Which of these does the RS-232 standard *not* specify?

    A Serial signalling protocol.
    B Types of connector.
    C Maximum frequency.
    D Allocation of connector pins.

5 A shift register:

    A converts parallel data to serial data.
    B moves bits from one register to the one next to it.
    C has three-state outputs.
    D converts serial data to parallel data.

6 In an even-parity system, which one of these bytes is in error?

    A 11001100.
    B 11110011.
    C 01101011.
    D 00110101.

7 An R-2R 'ladder' is used in a:

    A DAC.
    B flash converter.
    C successive approximation converter.
    D operational amplifier.

8 One of the disadvantages of a successive approximation converter is that:

    A it is very expensive.
    B the input must not change too fast.
    C its output is limited to 8 bits.
    D it is inaccurate.

6 Planning the system

Summary

The advantageous feature of digital circuits are described. Discussion of noise, including EMI, leads on to choosing a suitable logic family for building a project. Connections within families and between families are considered. The design of the circuit board is discussed, together with precautions that may be taken to ensure correct operation. Breadboarding techniques are outlined.

Digital systems

A digital system in general has fewer problems than an analogue system. One reason is that a digital system operates on binary data. Voltages are either high (if they exceed a certain minimum 'high' value) or they are low (if they are less than a certain maximum 'low' value). There is usually no difficulty in outputting voltages that are comfortably within the specified high or low ranges, and there is no difficulty in inputting these voltages and having them correctly interpreted.

The second advantageous feature of digital circuits is that most of them are driven by a system clock. We make sure that voltages have had time to settle and *then* the changing clock edge triggers off the next state of the system. Both voltage levels and time are accurately managed, so eliminating many of the uncertainties prevalent in analogue circuits.

Noise

Any unwanted electrical signals present in a circuit are referred to as *noise*. If the amplitude of the noise is great enough, it degrades the wanted signals. In logic circuits, it may cause a logic low level to be read as a logic high level or the other way about, so that data is corrupted. It may have a similar effect on control signals causing a device to reset, for example, or to latch its input at the wrong time. Similar effects can result from noise on the supply lines.

It is usually possible to minimise noise to an acceptable level but impossible to remove it altogether. Noise may be generated in the circuit itself, especially in resistances and semiconductors. It may be picked up from other parts of the same circuit, especially where tracks on the circuit board run side-by-side (crosstalk). It may also be picked up from outside as electromagnetic interference (EMI).

For the reasons stated above, minor variations in power supply, ambient temperature, and component values have little effect, if any, on the operations of a digital circuit. For the same reasons, *noise* in the form of voltage spikes is largely ignored by digital circuits. This is why digital circuits have become so widely used, even for applications such as audio recording and playback in which the original input and final output are necessarily analogue.

In spite of the essential robustness of digital circuits, there can be problems with these too. The problems are more severe when we try to push the circuits to the limits of their capabilities, particularly as we increase processing speeds to their maximum. In this chapter, we look at some of the problems that may arise with digital circuits in microelectronic systems and how we may try to overcome them.

Many of the problems arise because of two related reasons:

- The signals in microelectronic systems are mostly of *radio frequency*.

- The signals being of high frequency, their rise times and fall times are extremely short. Putting this the other way round, there are high *rates of change* of voltage and current.

A third source of difficulty is *electromagnetic interference* (EMI).

As a result of one of these factors, a logic gate may become set to the wrong state and the system will fail.

Noise immunity

Fig. 2.12 shows the input and output levels of LSTTL and CMOS gates. The voltage at a gate producing a high output is at least 2.7 V. A gate receiving that output will accept it as a high input if it is at least 2.0 V. This means that the gate will still give the correct result if there is a spike of, say −0.5 V on the output signal, taking it down to 2.2 V. It still could *just* work if the spike is −0.7 V, but might fail occasionally. These values show that the *noise immunity* of LSTTL is 0.7 V for logic high. For logic low the corresponding voltages are 0.5 V and 0.8V, giving a noise immunity of 0.3 V. Spikes as high as this

---

Electromagnetic interference

Any flow of electric current generates electromagnetic radiation (radio waves), the larger the current the stronger the radiation. When this radiation passes through a piece of electronic equipment it generates voltages in the signal lines and in the power lines. These may cause the circuit to operate incorrectly. EMI may pass directly as radiation or may be transmitted along the mains power lines.

There are many sources of EMI, one of the worst being switching currents on or off. The switching of the clock in a microelectronic system generates high frequency radio waves which may interfere with the operation of the system itself or of a neighbouring system. The switching of signals in the data and address busses has the same effect.

EMI can also arise from outside sources such as domestic equipment (switching the motors of refrigerators, washing machines), in industrial plant, and in the ignition systems of motor vehicles. In these examples the loads are inductive so a high voltage is built up as the switch contacts open. This causes arcing at the contacts, which is a powerful source of EMI.

Another source of EMI is the electric mains, where the 50 Hz (or 60 Hz in USA) may appear as ripple on the power lines of a microelectronic system. This can be eliminated with good shielding and a properly designed power supply.

Natural sources of EMI include thunderstorms and the effects of cosmic radiation from outer space.

may occur quite often in a badly designed circuit.

The noise immunity of CMOS operating at the same voltage (5 V) is 1.45 V at both high and low logic levels. This is better than LSTTL. Increasing the supply voltage to 15 V increases the noise immunity still further, to 3.95 V at high and low levels. For a system working in a noisy environment, it may pay to use CMOS operating at 15 V.

As mentioned in Chapter 2, the noise immunity of ECL is less than that of TTL or CMOS, making system design with this family a difficult task.

Logic and clock rate selection

One of the first decisions to be made, after deciding on the processor (Chapter 4), is what IC family is to be the main one used. The choice is between:

- 74LSXX series (low power Schottky TTL). A wide range of ICs available, but must have a regulated 5 V supply. The original 74XX series is now available only in a restricted range of types. There are other TTL sub-families for special applications.

Schottky: Schottky diodes are incorporated in the gate circuits of the 74LSXX family. They act to prevent the transistors from becoming saturated when they are switched on. This means that the transistors can be switched off more rapidly, so reducing the propagation delay of the gate.

- CMOS 4000 series. A wide range of ICs is available, including many complex ones. It operates on 3 V to 15 V supply. It has a large fanout and good noise immunity. CMOS is slower than TTL but fast enough for most applications.

- CMOS versions of TTL series, the 74HCXX and 74HCTXX sub-families. They have greater noise immunity than regular TTL.

- ECL is fast but has low noise immunity, and circuits are troublesome to design.

Another choice to be made is the speed of the system clock. A safe general rule is to make the clock as slow as possible. In addition, use the slowest family that will do the job. Usually this means settling on the CMOS 4000 series. The reason for keeping the system speed as low as possible is that this minimises EMI emission, which may affect the operation of nearby circuits and perhaps other sections of the same circuit. Conversely, low speed often makes the system less susceptible to noise from adjacent circuits and other sources of EMI.

Sometimes the choice of family is determined by what functions are available. For example, if long counter chains are needed, the 4000 series is the best. For octal buffers and latches, the 74LSXX series has more types to offer (Chapter 5). On occasions, it happens that the two functions required are not both available in the same family. In such cases, it is necessary to use an interface between them. The supply voltage must be compatible with both families, which usually means operating at 5 V.

When connnecting gates and other logic inputs and outputs, remember that there is a limit to the amount of current that an output can provide in the logic high state. There is also a limit (usually larger) in the amount of current a gate output can *sink* is the low logic state. There are also limits on the amounts of current to be sourced or sunk to make an input register as low or high. The result is that there are definite numbers of gates that can be reliably driven by the output from a given type of gate. This number is known as the *fanout*. The table below shows the fanout of various combinations of families.

Driving output	Driven input	
	74LSXX	All CMOS
74LSXX	20	*
74LS buffers	60	*
CMOS4000	1	50
74HCXX and 74HCTXX	10	*
74HC/HCT buffers	15	*

* a virtually unlimited number of gates.

It can be seen that, except for a CMOS4000 output driving a 74LSXX input, fanouts are sufficiently great for most designs. A problem arises with certain microprocessors that have low fanout. Buffer ICs may be needed to prevent the bus from being overloaded.

In the case of TTL gates (which includes gates of the 74LSXX series) the minimum output voltage (2.7 V) that counts as logic 1 is less than the minimum logic 1 input voltage (3.5 V) to a CMOS gate (or 74HCXX gate) running on the same +5 V supply. This is shown in Figure 2.12. A high output sent from a TTL output to a CMOS (or 74HCXX) input may fail to be recognised as high. Because of this, any connection between a TTL output and a CMOS (or 74HCXX) input needs a pull-up resistor of 2.2 kΩ, wired to the supply line. Alternatively, substitute a 74HCTXX device for the CMOS device as the input and output voltages of the HCT series are compatible with TTL.

Many chips, including RAMs and some CPUs, are made using NMOS technology. NMOS is compatible with CMOS, so what has been said about fanout of CMOS also applies to NMOS.

When a gate changes state, it produces a change in the amount of current drawn by the gate and by any unit being driven by it. The result is a sudden change in voltage on the power lines. In other words, there is *noise*. The situation is worse with synchronised logic in which many gates change state at the same time. Conversely, devices such as flip-flops and counters are particularly sensitive to noise. Noise may cause them to change state or mis-count.

Noise on the 0V (ground) line in minimised if the line has low impedance. Any suddenly increased current flowing into the line is immediately carried away to the 0 V terminal of the power supply. To reduce this noise we can:

- Make power tracks wider, particularly the 0 V track. If possible, widen the 0V track into a *ground plane,* occupying all vacant areas of the board .

- Make power tracks as short as possible. This advice applies to tracks of *all* kinds for, the shorter the track, the less likely it is to pick up EMI, and the less likely it is to emit EMI. This is particularly necessary when designing the system clock, a potent source of noise and at the same time easily affected by noise.

- Decouple the supply. This consists of placing capacitors (for example, 100 nF multilayer or disk capacitors) across the supply lines, close to the IC which is to be decoupled. The distance between the capacitor and the terminal pins must be kept as short as possible and so must the leads of the capacitor. Otherwise the

---

Track resistance

The resistance of a typical circuit board track is 4 mΩ per cm. If a section of track is 10 cm long its resistance is 40 mΩ. If one end is connected to the 0 V terminal of the power supply and a current of 100 mA is applied at the other, the voltage difference betrween the track ends is 100 mA × 40 mΩ = 4 mV. This may not be enough to affect the operation of a logic gate powered from the track, but the voltage spike will be greater and more effective if:

- the current is greater.
- the track is longer.
- the track is narrower.

inductance of the track and leads in combination with the capacitor produces a resonant circuit. This will cause 'ringing' whenever there it a change of logic level, with unpredictable results. For the very fastest logic, there are IC sockets which include the capacitor directly under the IC. For simpler, slower, circuits there is less problem with decoupling. A few 100 nF capacitors should be scattered around the board, one capacitor to every 5 or 6 ICs. It is also a good idea to decouple the supply with a larger capacitor (say a 47μF electrolytic) where the supply lines enter the board.

Ringing: When there is a change of logic level from 0 to 1, the voltage rises to 1, then overshoots, then oscillates for a while around 1 with gradually decreasing amplitude until it settles at 1.

Crosstalk

Another major problem is *crosstalk*, which is the picking up of a signal by an adjacent track. This is particularly serious in digital circuits because the signals have rapid rise and fall times. Changes in level are equivalent to a very high frequency signal. Where two tracks lie side-by-side for a few centimetres there is capacitance between them. At very high frequency the impedance of a capacitor is reduced to a minimum. Signals pass freely across from one track to the next.

To avoid this effect we can:

* increase track spacing.

* keep parallel runs of tracks as short as possible.

* avoid running signal tracks parallel to power line tracks (large voltage fluctuations on power lines could swamp the signals).

* keep tracks carrying input signals away from tracks carrying output signals. This applies to the board as a whole (keep input circuits away from output circuits) and to individual ICs (keep tracks leading to input pins away from tracks leading from output pins).

Transmission lines

When very high frequency signals are used in a microelectronic system the wavelength of the signals may be comparable to the lengths of some of the tracks. Then the normal rules of conduction no longer apply. Instead the tracks must be treated as *transmission lines*. A transmisson line has a *characteristic impedance* depending on its dimensions and other factors. If the output impedance of a signal source and the input impedance of the receiver both match the characteristic impedance of the line, the signal is transmitted along the line with virtually no loss. If they do not match, the signal may not reach the far end, or may be received with much reduced power. A portion of it may be reflected back from the receiving end causing standing waves in the line and also resulting in ringing and other effects. These make the operation of the computer very erratic. The design of systems involving transmission lines is outside the scope of this book.

Design points

Here are a number of design points which will improve the performance of a microelectronic system:

Unused inputs: It may often happen that one or more inputs to a gate or logic device are left unused. Sometimes a chip contains more gates than are required, so there are whole gates unused. There are rules about what to do with unused gate inputs, and other unused inputs such as resets, presets, and chip enable inputs. The rule is simple with CMOS. All inputs must be connected to the positive supply line, the negative supply line or the output of another gate. You must always do this, even when you are breadboarding an experimental circuit and are using only one gate out of the four on the chip. If the inputs of the other three are left unconnected, the gate you are using may not behave correctly and may take a large amount of current. A rule related to this is that all connections to the power line must be made *before* the power is switched on. Conversely, the power supply must be switched off *before* any alterations are made to the circuit.

With TTL, inputs of gates or other circuits that are left unconnected act as if they have a logic high input. However, if a gate that is being used has spare inputs, it is better not to leave these unconnected. Unconnected inputs reduce noise immunity. Instead, connect unused inputs to:

- the positive supply, through a 1 kΩ resistor.
- the 0 V line (no resistor required).
- one of the used other inputs of the same gate.

Fig. 2.10 shows two examples of this technique.

Debouncing inputs: When a switch or key closes, the unevenness of the contact surfaces (on a microscopic scale) cause it to close and open again several times before it eventually closes permanently. We say that there is *contact bounce*. The rapid succession of on-off states is not noticeable when, for example, we switch on a lamp or a motor. But a logic gate reponds so rapidly to changes in input that it detects every closure and opening. If, for example, a key is intended to send one pulse to a counter circuit every time it is pressed, the counter may register five or more 'presses' each time it is pressed once.

There are times when contact bounce does not matter. For instance, when we reset a flip-flop, it is reset on the first contact and repeating the action half-a-dozen times has no further effect. When contact bounce does matter we use special circuits to debounce the switch. In Fig. 6.1 the capacitor slows down the rate of change of logic level so that the noise from the switch is absorbed. This technique is effective but may make the action of the switch too slow. If a large-value capacitor is used to thoroughly debounce the switch, it takes longer to

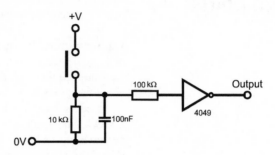

Figure 6.1 *This debouncing circuit gives a low output when the push-button is pressed. This is a suitable output for triggering an active-low input.*

charge and discharge. Switching on and off can not be repeated as rapidly as required. In this case the logic of Fig. 6.2 is be used. The set-reset flip-flop is made to change state by bringing one or the other of its inputs low. The change over occurs the *first* time the switch makes contact and the flip-flop does not change state again during the subsequent bounces.

In Chapter 12 wew discuss how to debounce a switch by using software instead of hardware.

Watchdog timer: In spite of all the precautions, it may still happen that a bit in memory or in one of the registers of the CPU may become

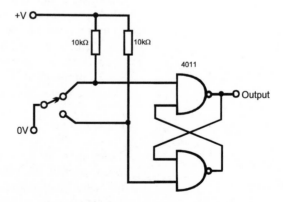

Figure 6.2 *The absence of capacitors in this debouncing circuit mean that it responds instantly to a change in the position of the switch.*

altered by electromagnetic means. Substituting a 0 for a 1 or 1 for a 0 can provide the CPU with false data or, worse still, may make it jump from its present place in the program to some unpredictable place in another part of RAM, where there may be no program. The result may be errors in output or the program may crash completely.

The watchdog timer is a technique for avoiding such dangers. It keeps a watch on the CPU, making sure that it is operating correctly. The timer can be an ordinary electronic timer circuit, such as one based on the popular 555 or 7555 timer IC. Or it can be a timer built in to the processor. Some I/O ICs also have one or more timers included.

The principle of the technique is very similar to the routine followed by a night watchman, who has to report to a supervisor or automatic security system at frequent intervals while on his rounds. If he fails to check in at the expected time, it indicates that something is wrong and action is taken. Similarly, the CPU is programmed to trigger the timer at frequent intervals, perhaps every millisecond. The period of the timer is a little longer than the interval at which it is triggered. When the timer has been triggered, its output goes high and it should be triggered again before the period is over. The output of the timer is connected to the reset input of the processor and, as long as the timer holds this high, the processor runs normally. If a fault develops and the CPU is no longer following the program correctly, it no longer triggers the timer. After a millisecond or so, the timer output falls to logic low. This automatically resets the CPU, which jumps back to the beginning of the program and starts again.

Breadboards

Most if not all of the microelectronic systems you build, program and test in your practical classes will be assembled on a breadboard. From what has been said about the problems of digital circuits, it might be wondered if you will ever get your breadboarded circuits to work. There is not much need to worry. The systems you will build have a clock of relatively low frequency so that there is little EMI. The slow action also gives the circuits time to respond and settle. Connecting wires are short (always use wires as short as possible on the breadboard). Also your systems are small, rarely consisting of more than three or so ICs.

One of the more serious problems with using a breadboard for microelectronics systems is the large number of connecting wires that is sometimes required. The situation is worse if the circuit includes busses. Inserting the wires systematically is easy, but problems arise if the circuit fails to work correctly. It then becomes difficult to follow the connections among a cluster of wires. Worse, if a few of the wires accidentally come out of their sockets, it is sometimes difficult to know where to replace them without carefully checking through the whole system. If this seems likely to happen in your project, it is worth considering building at least part of the circuit on a rectangle of stripboard. Mount all ICs in sockets, so that they may easily be removed for testing the connections.

Questions on planning the system

1 What is meant by noise in electronic systems? How can it be reduced?

2 What is noise immunity? Explain why CMOS has good noise immunity.

3 Discuss the points that have to be considered when deciding on what logic IC family to use in a microelectronics project.

4 Explain, with examples, the meaning of the term *fanout*.

5 What is *crosstalk* and how may it be avoided?

6 What is meant by *contact bounce*? Describe two circuits which reduce or eliminate it.

7 Describe the function of a watchdog timer.

Part B – The Software

7   Instructions

Summary

The CPU reads a program in a series of fetch-execute cycles in which it reads a machine code instruction and then obeys it. Writing programs in machine code is difficult, but easier if we use an assembler program and write it in mnemonics. The assembler turns the mnemonics into machine code. A program written in assembler can have a title and comments added to it, to make it even easier to understand.

A CPU can do many different things, but it does nothing unless it is told to. It needs to be issued with *instructions*. The instructions are issued to it in the form of a *program*, stored in memory. The instructions that a given CPU can understand and obey are its *instruction set*.

Different types of CPU have a different instruction set; that is, they have a particular collection of instructions suited to the architecture of the CPU. Some kinds of CPU have a large instruction set with several hundred instructions in it (CISC). Others work on a much smaller set, usually fewer than a hundred (RISC). Usually all the members of a *family* of CPUs have the same instruction set. Some members of the family, especially the newer ones, may have a few additional instructions in their set.

The CPU obtains its instructions by going to an address in memory and reading the instructions stored there. There are very fast CPUs that read more than one at a time but most CPUs read one instruction at a time and act on it before fetching the next one. This is the type of CPU we describe in this chapter. We shall also confine our descriptions to systems in which the data is operated on as single bytes.

Fetch-execute cycle

When it is operating, a CPU repeatedly goes through a cycle known as the *fetch-execute cycle*. As its name implies, this cycle has two stages:

- The CPU fetches an instruction from memory.
- The CPU executes it, that is, it obeys the instruction.

The cycles are repeated continuously as the CPU works its way from the beginning of the program to the end.

The instruction is a byte of data stored in memory. Reading a byte from memory takes place in ten stages:

- The address of the byte to be read is transferred along the internal bus of the CPU to the address bus register.
- The address is placed on the address bus (Fig. 7.1).
- The address is decoded by the logic (partly on the memory chip), to select the location for reading.
- Reading is enabled. In Fig. 7.1 the $\overline{\text{WRITE ENABLE}}$ line is already high so reading is already enabled.
- The $\overline{\text{CHIP ENABLE}}$ line goes low, to switch the memory chip outputs from the high-impedance state to a low-impedance state (outputs high or low).
- Data from the memory location appears on the data bus.
- The CPU stores the data from the bus in its data bus register.
- The data is transferred to the internal bus of the CPU.
- If the data is an instruction, it is stored in the instruction register.
- While the last two states are occurring, the $\overline{\text{CE}}$ line goes high and the data is no longer present on the data bus.

Assuming that the CPU has only one read-write enable control line.

Assuming that the data is an instruction, the logic of the control unit causes one of many different actions to occur.

Example:

The data register of an 8085 CPU holds the following byte:

01001111

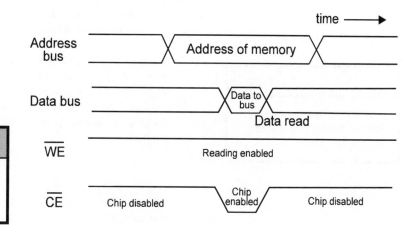

Figure 7.1 *Reading a byte from memory involves a sequence of logic levels on two lines of the control bus. The* WRITE ENABLE *line stays high as this is a read operation. Data from memory is put on the bus when the* CHIP ENABLE *line goes low.*

Test your knowledge 7.1

Why is the address put on the bus before the chip is enabled?

Remember that data transfer is by *copying*. The data now in the instruction register is still stored in memory, where it can be read as many times as the program requires.

The latches of the register are either set (= 1) or reset (= 0). When the control unit receives this data it copies to register C (a general-purpose register) the data that is in register A (the accumulator). This is an internal movement of data so a single byte is enough to tell the control unit what to do. How the control logic works, that is to say, how the input of 01001111 makes the control logic take the action we have described, is outside the scope of this book.

Example:

At another time, the data register holds:

00110111

This instructs the control unit to set the carry bit of the status register to '1'. As above, a single byte is all that is necessary for this instruction. It is obvious that, if we always had to write out data as 8-bit binary numbers, as above, it would be all too easy to make mistakes. Instead, we express the numbers in hexadecimal. So the two instructions described above become:

4F, for copying data from A to C, and
37, for setting the carry bit

Some operations require two bytes. The first byte states what is to be done. On loading this byte, the CPU loads the next byte in memory to find out what value to operate on.

Example: The code E6 (actually 11100110) tells the control unit to AND register A with the binary value that is in the next byte. The control unit has to send the CPU back to memory to find out the value of this byte, which is stored in the next location to the instruction E6. Therefore, the full instruction might be:

E6 4A

This tells the control unit to AND the content of register A with the value 4A.

Test your knowledge 7.2

What is the code to make the CPU AND the contents of its register A with the value $2B?

Writing to memory

A sequence of operations may often end in the storing of a result in memory. There are opcodes for storing the contents of various registers. The operand is the address at which the data is to be stored. In a 16-bit system, it takes two bytes to specify the address. This means that the opcode is followed by two bytes, making the instruction three bytes long. The sequence of storing data is almost the reverse of the sequence for reading data. It has nine stages:

- The address to which the byte is to be written is transferred along the internal bus of the CPU to the address bus register.
- The address is placed on the address bus (Fig. 7.2).
- The address is decoded by the logic (partly on the memory chip), to select the location for writing to.
- The data from the register (often the accumulator) is placed on the internal bus.
- The data is latched into the data bus buffer.
- The data appears on the data bus (Fig. 7.2).
- Writing is enabled. In Fig. 7.2, the $\overline{\text{WRITE ENABLE}}$ line is made low.
- The $\overline{\text{CHIP ENABLE}}$ line goes low, to store the data from the bus in the addressed location.
- The $\overline{\text{CE}}$ line goes high, followed by the $\overline{\text{WE}}$ line. The address and data are no longer present on their busses.

The instructions for a program are stored in memory (RAM or ROM) as a sequence of consecutive bytes, which are the opcodes, and operands of a program. This is known as *machine code*. It is the only form of program on which the machine (the CPU) can work. If the

Remember that in this book a hex value is prefixed by $.

Test your knowledge 7.3

What is the result of ORing the two values quoted in the box?

Bitwise logic

CPU logic is always done *bitwise*, that is, corresponding bits in two values are subjected to the logical operation.

Example: Suppose that, as the result of a previous operation, register A holds the value $AC. To AND this bitwise with $4A we must first write out the values in binary:

10101100
01001010

The rules for the AND operation are:

$0 \bullet 0 = 0$
$0 \bullet 1 = 0$
$1 \bullet 0 = 0$
$1 \bullet 1 = 1$

Note that in these equations '•' represents the operator 'AND'. The result on the right is true (= 1) only if *both* bits on the left are 1.

In the example, only the fourth pair of bits from the right is 1•1. The result of ANDing is:

00001000

In hex, this is $08, which is the value that would be left in register A at the end of the operation.

Opcodes and operands

The first byte of an instruction tells the CPU what type of operation to perform. It is known as the *operational code*, or *opcode.*

Sometimes the opcode is followed by one or two more bytes, which give the CPU something on which to operate. It may be a value to work on, or it may be an address where a value will be found stored. This byte or pair of bytes is known as the *operand.*

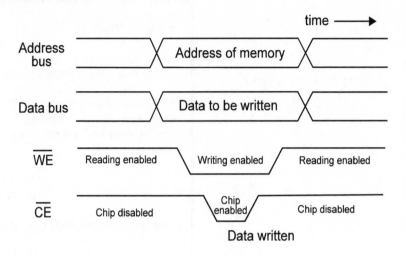

Figure 7.2 *The timing of signals on the busses and control lines must allow logic levels to settle and for logical responses to be completed. The sequences illustrated here and in Fig. 7.1 normally take at least 3 or 4 cycles of the system clock, possibly up to 12. If the system clock is running at 100 MHz, four cycles take 40 ns.*

program is first written in machine code, this is also known as the *source code*.

Assembler

Writing in machine code is not easy. Fortunately, there is software that makes it possible to write the source code in other forms and have it automatically turned into machine code. The simplest type of software is *assembler*. It is much easier to write in assembler than in machine code. One reason is that we write assembler in *mnemonics* instead of in opcodes. A mnemonic is a group of up to four letters to represent each opcode and there is a different mnemonic for each opcode. The point of mnemonics is that the letters help us remember what the operation does.

> *Example*: The instruction 00110111 on the data bus is the equivalent of $37. The opcode $37 tells the CPU to set the carry bit in the status register. If we are programming the system in assembler, the corresponding mnemonic is STC. This is easier to remember because the three letters remind us of the action: 'SeT Carry'. So we type in STC and the assembler program converts this to the opcode '37' and adds it to the program.

Different assemblers may use slightly different mnemonics but the

---

Learning assembler

The only way to learn how to use an assembler is to practise writing programs. There are many suggestions for programming exercises in this book. Many different microcontrollers are suitable for this work and each has its own assembler program. The example in this section is based on the assembler for one of them, the Atmel '1200' microcontroller.

The assembler you use may differ in detail from the examples given here. However, the examples used in this chapter are short and they are simple in structure, so it is easy for you to find the equivalents on your assembler.

---

principle is the same for all. Some of the mnemonics may be identical in different assemblers. An assembler is written with a particular CPU in mind. It has, for example, to take into account the widths and numbers of registers that the CPU has, and whether or not they can be used for arithmetical and logic operations. Some CPUs can do things that others can not do. In such cases, the assembler may include special mnemonics for managing these particular actions.

Although an assembler simplifies program writing by using mnemonics, it still requires us to give the CPU its instructions systematically.

Example:

Note that this is part of an imaginary program, so do not worry about what it is supposed to be doing

Here is a short section of program written in the assembler of the '1200' microcontroller:

```
ldi r17,37
inc r17
mov r5,r17
```

The mnemonic 'ldi' is short for 'LoaD Immediate', which means load the value which follows immediately.

In the '1200' assembler, numbers prefixed with 'r' refer to one of the CPUs 32 registers. Values are taken to be decimal unless prefixed by the $ symbol. Given this information, it is easy to see that this sequence means:

Load register 17 with the value 37.
Increment register 17.
Move (or copy) to register 5 the content of register 17.

With the assembler program running in a PC or similar computer, the program listing given above is typed in. It appears on the computer screen just as shown above. In this book, we use a `different style of type` to indicate programs and other screen displays. This style is the same as used by our assembler and many others to display text on the screen.

Another advantage of assembler is that we can use decimal numbers instead of having to convert every value into hex. The assembler does all the converting when it turns our source code program into machine code.

When the program is typed in, it appears in a window on the screen. If it appears to be correct, clicking on the 'Assemble' button at the top of the screen causes the program to assemble it, that is, to turn it into machine code. This is much quicker than looking up a table of opcodes. The machine code is stored in a file, with the extension *.obj.*

At this stage, instead of assembling, the assembler may report that there are errors in the program. It states the number of line or lines on which the errors occur and gives a brief description of the type of error.

Example:

The first version of this program used register 2 where it now uses register 17. On assembling, an error was reported in line (1). This was because only registers 16 to 31 can be used for arithmetical operations, including loading numerical values. After the program had been edited to change 'r2' to 'r17', assembly proceeded without error.

When a program is being typed in, it appears in a special window on the screen. It is possible to have other windows open on the screen at the same time. One useful window is the Processor window. This displays the state of the program counter, the stack pointer, the flags and other useful data. Another window, of special use in checking the short example program above, is the Register window. This shows the content of all 32 registers. It makes it simple to follow the changes in the values held in registers as the program is run. If you simply 'run' the program by clicking on the 'run' icon in the toolbar, the program runs so fast that it is impossible to keep track of all the changes. Instead, click on the 'single step' icon. Then the program runs one line each time you click, and you can look at the results of each step. You can see the changes in the values stored in the registers and, at the same time, watch for changes in the program counter and in the flags in the status register. This is another big advantage of using an assembler program.

The programs described in this section are being written on a PC or other computer, using assembler. When complete and assembled, they may be saved on to disk, downloaded into the EPROM of a microcontroller or 'burnt' into a PROM, using a PROM programmer.

Example:

The easiest way to follow the changes is to run through the program one line at a time. At the start of the program, all registers hold zero. After the first line, r17 holds '25'. (although *we* are working in decimal, the *assembler* is working in hex). After the second line, r17 changes to '26'. After the third line, r17 still holds '26' and the same value is copied to r5.

At the same time, the program counter in the Processor window begins at 0, then changes to 1, then to 2, and then to 3. There are no changes in the flags.

As a reminder that the essential thing about an assembler is that it assembles machine code for us, it is possible to display the actual machine code of the short program. The memory of the '1200' is 16 bits wide, so it can store both the opcode and the operand in a single 16-bit *word*. The numbered memory locations are listed on the left, and in the next column, we see the machine code. The original assembler listing is repeated on the right.

Errors

There are three main kinds of programming error:

• Telling the CPU to do something it can not do.
• Telling the CPU to do something it can not understand.
• Telling the CPU to do something it can understand and do, but which produces a result you did not intend.

The assembler (and most other software for producing programs) will report the first two kinds of error. An example of the first kind is mentioned in the text. An example of the second kind is if you make a typing error, typing ICN instead of INC. As another example, you may follow the opcode with too few or too many operands. These errors are often reported as *syntax errors*.

Errors of the third kind are the programmer's responsibility and are not reported. Provided that the CPU can do what you tell it, it will do it. This is why you need to debug your programs thoroughly. The assembler program may include a section for debugging, or you may use a separate debugging program.

Example:

The display of memory content of the example program looks like this:

```
+00000000:    E215      LDI   R17,0x25
+00000001:    9513      INC   R17
+00000002:    2E51      MOV   R5,R17
```

This display reveals that the program occupies the first three 16-bit memory locations, which contain E215, 9513, 2E51, in that order. We shall not be looking at the machine code after this.

On the screen, there is a further entry on the extreme right of the first line:

```
;    0x25 = 0b00100101 = 37
```

This is a comment or remark put there by the assembler. It explains that the hex value 25, corresponds to the binary value 00100101 and to the decimal value 37. When the assembler lists a program or makes comments, a hex value begins with 0x. A binary value begins with 0b and a decimal value is written as we would normally write it. Other assemblers may set out values differently.

Note the semicolon preceding the comment. This is a conventional way of indicating that what follows on that line is not part of the program. We shall return to this point in a moment.

Whether or not the assembler that you are using has exactly the same features as the one described here, the point is that an assembler allows you to type in a program and assemble it. It searches for and reports errors. It allows you to view the program in both assembler and in machine code. What it does not do is inform you if what you intended the CPU to do, is not the same as what you have actually told it to do. This is where debugging is important.

Improving the program

The bare program as listed earlier is all that is essential, but can be made more understandable to the programmer and to others who may want to study it and perhaps modify or extend it. This is done by adding a title and some comments or remarks. These are preceded by a semi-colon. When assembling, anything between the semi-colon and the end of the line is ignored by the assembler. However, it may be of great interest to programmers and others.

Example

This is one way in which the sample program can be improved by a title and comments:

```
; Program example 1, introducing assembler

ldi r17, 37    ; loads register 17 with
               ; blood temperature.
inc r17        ; increments temperature.
mov r5,r17     ; stores incremented temp in
               ; base register.
```

Note the positions of the semi-colons. The title explains what the program is about. The comments on the right explain what happens at each step. Long comments can be carried on to the next line (after a semi-colon).

Comments are important because an uncommented program of more than a few lines is extremely difficult to understand, even by the person who wrote it, and especially a few weeks after it is written. Writing useful comments is an acquired skill. They must not be so brief as to be unintelligible. They must not be so long that the actual program becomes 'lost' in a mass of multi-line comments.

Another advantage of assembler compared with machine code is that assembler allows us to use *labels*. A label is a word used to identify a:

- program line.
- register.
- variable (number).

Using labels makes it much easier to understand what a program is doing and how the different parts of it link together. Labels must be defined at the beginning of the program.

Example:

```
; Program example 1, introducing assembler

.def counter  = r17
.def trial = r5

.equ body = 37

ldi counter, body  ; loads r17 with
                   ; body temp.
inc counter        ; increment temp.
mov trial, counter ; store in trial.
```

The statments beginning with .def and .equ are not *instructions* for the CPU and will not appear in the assembled program. They are *directives,* which tell the *assembler* what to do. For instance, the first directive tells the assembler to substitute the register number r1 every time it comes across the word 'counter'.

After the title, but before the actual program listing, are two *definitions* (.def) which tell the assembler the names we are giving to two of the CPUs registers. In the program, we always refer to the register by its name instead of its number. There is also a statement to say that the label body is equal (.equ) to 37. This is actually an extra step that was not included in the previous program. There we directly loaded counter with 37. Now we are loading 37 into a constant called body. After that, we copy the value of body into counter. The name body can be used anywhere later in the program where we need to use the value of body temperature. In addition, if for any reason we want to change the value throughout the program, we can simply change it in the .equ definition.

Rarely does a program run straight through from beginning to end. More often, the CPU is required to jump from one part of the program to another, often skipping back to repeat a part of the program several times. In machine code, there are instructions to tell the CPU to jump, and giving it the address in memory to jump to. This is another cause of difficulty. Calculating the exact number of bytes to jump over, and expressing that number in hex is a frequent source of error. If the program is subsequently altered, perhaps by inserting extra sections of code, many of the 'jump to' addresses will be wrong. It is then necessary to work through the program correcting all the addresses. There is the further point that an address itself (example $E28A) has little meaning to the programmer or to someone else reading the program. Fortunately, an assembler allows us to label the lines of the program, so that the CPU can be told to jump to this named line. Even if we insert new pieces of program, the label stays with its line and there is no need to correct it.

Example:

The sample program processes the value stored in trial, then has to come back to start work on the next (incremented) value. It needs to jump back to the line in which the value in count is incremented. We place a label again at the beginning of this line to indicate the line to which the CPU is to jump back.

```
;Program example 1, introducing assembler

.def counter = r17
.def trial = r5

.equ body = 37

        ldi counter,body ; loads r17 with
                         ; body temp.
again:  inc counter      ; incrementtemp.
        mov trial,counter; store in trial.
```

This short section of program does not include the line later in the program which tells the CPU to jump back to again.

A dry run

In the early stages of writing a program or a section of a program it often helps to take paper and pencil and work out how the values in the registers and in memory will change, line-by-line.

This is known as a *dry run.* The same kind of tryout can be run using the assembler program or some of the high-level languages, but things may seem clearer to you on paper than on the screen. In addition, it is possible to cross out and scribble comments on paper, but less easy to do this on the screen.

Here is a dry run of the assembler program:

	r5	r17
Start	0	0
ldi r17,37	0	37
inc r17	0	38
mov r5,r17	38	38

The columns show the contents of the registers *after* each line has been run.

Later in the program, when the CPU has finished working on the current value of trial, it is told to jump back to again. Then trial is incremented and the CPU does the processing again, using the new value.

This completes our account of the principles of assembler programs. Summing up, its main advantages are:

- Easily remembered mnemonics.
- Can use decimal numbers.
- Comments.
- Automatic assembly.
- Error reporting.
- Labels for addresses, registers, and variables.

There is more to be learned about assembler, but this is best done by working the practical examples in the Activity sections below.

Activity 7.1 Starting assembler

You need:

- A computer with assembler software installed.
- The manual for the assembler (assuming that you already know how to use the computer).

Find out how to enter programs into the assembler. Try out any working examples for beginners that are explained in the manual. You are not expected to memorise or to use all of the instruction mnemonics. As you work these activities, write on a piece of card the mnemonics and directives that you actually use. This will give you a quick reference list of the most useful (to you) of the mnemonics and directives.

Look for the mnemonics and directives that most closely match the ones we have used in the example program above. Use these to write a version of our program that will make your chosen microprocessor or controller perform the same task. The set of registers you have available may be similar to those of our '1200' microcontroller, but may be different. If they are different, remember that the aim of the program is to get the CPU to load a number, to increment it, and to store it in another register. If you do not have enough registers for this, store it at an address in RAM.

Assemble the program you have written and correct any errors that you have made.

Run your program, single-stepping it, and watching the changes that occur in the register, the program counter, and the status register.

Activity 7.2 Adapting the program

You need:

- A computer with assembler software installed.
- A user manual for the assembler.
- Your card of frequently used mnemonics.

If you have not already done so, improve the program you wrote in Activity 7.1 by adding a title and comments.

Re-write the program to make the CPU do something slightly different. For example, make it load a different number. Make it store the incremented number in a different register. Make it increment the number twice before storing it. Make it decrement the number instead of incrementing it. Using the mnemonics from your version of our program and perhaps one or two others from your instruction set, there are many different ways of programming the CPU to do something slighty different. Add comments to each short program you write, so that others will be able to understand what it is supposed to do.

Problems on instructions

1 What is a fetch-execute cycle? Give examples based on a CPU that you are studying.

2 Describe a read cycle, illustrating it by a timing diagram (not necessarily to scale) for the CPU you are studying.

3 Describe a write cycle, illustrating it by a timing diagram (not necessarily to scale) for the CPU you are studying.

4 What is machine code? Give examples of a few opcodes and explain the operands that they require.

5 What is an assembler program and what are its advantages over writing a program in machine code?

6 List a short assembler program that you have written. Explain concisely what the CPU does at each stage.

7 Do a dry run of the program you described in your answer to Question 6.

8   High-level languages

Summary

A high-level language program produces machine code from a program written in a form that is close to ordinary English. The many small steps of machine code and assembler are replaced by keywords that call up ready-made routines. Some languages use an interpreter, and others use a compiler. Two widely used high-level languages suitable for programming microelectronic devices are BASIC and C. Ladder logic is often used with PLCs.

Just as assembler is one stage removed from machine code, so the high-level languages are one stage removed from assembler. There are many high-level languages, a few of them very widely used, and several that have more specialist applications. CPUs can not understand high level languages, but there are several programs that allow us to write a program in a high-level language and then have it turned into machine code. In this chapter we look at three such programs, using C, BASIC and ladder logic.

The main difference between assembler and high-level languages is that assembler is related to machine code on a one-to-one basis. Each line of an assembler program corresponds to one fetch-execute cycle of the CPU. Because a CPU operates in short simple steps, programming in assembler can be very tedious and repetitive. With a high-level language, a single command may be the equivalent of several tens of fetch-execute cycles.

Example:

A program written in C might contain this line:

```
area = length * width;
```

The symbol * means 'multiplied by'.

Two variables, labelled `length` and `width` are to be multiplied together to obtain the value of a third variable labelled `area`. This command sends the CPU into a long sequence of fetch-execute cycles in which it loads the variables `length` and `width`, multiplies them together (a fairly complicated operation for a CPU that can only add or subtract) and finally store the result as `area`.

Variable: a data item which may have its value changed possibly many times while the program is running. There are several types of variable, including integer variables.

A high-level language program includes an extensive range of ready-made machine code routines for programming complicated sequences (such as the multiplication routine mentioned above) so that the programmer does not have to work them out every time they are needed.

Although programming in assembler may be complex and tedious at times, it does have the advantage that the user is always fully aware of what the CPU is doing. This can lead to efficient programming carefully tailored to the project in hand. The ready-made routines of a high-level language save much work, but are not necessarily efficient. In order to make them usable in a wide range of programming situations, they may be longer (and therefore slower) than an equivalent routine specially written to fit into one particular program. If maximum speed is essential, a programmer may use a high-level language for most of the program but use assembler when writing routines that need to run at maximum speed.

To be able to create a *source* program in a high-level language, you need the appropriate *language* program. This is the program that accepts the statements you type into your computer, and displays them on the screen. It allows you to edit your program, to check it for errors and to save it.

The form a program is saved in varies with the language. Most versions of BASIC, for example, use an *interpreter* program. The BASIC language program saves your source program not as machine code but as a sequence of bytes. Some of the bytes represent BASIC commands while others represent numbers and text. The numbers and text are represented by ASCII code. In this form, the program is meaningless to a CPU, which understands only opcodes and operands. To be understood by the CPU, the BASIC source program must first be *interpreted*. When you run the source program the *interpreter* takes over. It goes through the source program line by line, and command by

---

ASCII code

This is the *American Standard Code for Information Interchange* and is the standard way of coding letters, numerals and punctuation and also a number of printer-control commands. It is an 8-bit binary code with a range of values from 0 to 255. The first 32 values are the control characters, such as 'carriage return', 'bell' (produces a beep in modern computers but used to ring an actual bell in the old Teletype machines), and 'end of file'. Code 32 is a space. Codes 48 to 57 are the numerals 0 to 9, then there are some punctuation marks. Capital letters of the alphabet begin at code 65 and lower-case letters at 97. The codes in between these blocks are used for punctuation.

---

command and turns it into machine code. It sends this to the CPU. The CPU has to do four things at once:

- Read a line of the source program.
- Use the interpreter program to find out what the line means.
- Turn it into machine code.
- Obey the machine code.

The need to perform four tasks simultaneously makes running an interpreted BASIC program relatively slow. In addition, it is wasteful for the CPU to have to interpret a line every time it is used. Imagine a program in which the same sequence of calculations is repeated 100 times, with the CPU jumping back to the start every time it finishes. The lines of that section of the program are interpreted 100 times even though they always produce the same result. It would be far quicker if the lines were interpreted just once. This is the function of a *compiler*.

A *compiler* is a program that takes a program written in a high-level language and converts it directly into machine code. It does this in just one session and the machine code is stored in memory or on a disk. When the compiled program is run, the CPU has only to read the machine code and obey it. This is much faster than having to interpret every line as it comes to it.

C is a compiled language. There are also compiled versions of BASIC. Compiling is a complicated process that takes the CPU an appreciable time. Unfortunately, the program can not be tested and debugged until *after* it has been compiled. If any errors are found, the programmer has to go back to the source program, correct the errors, recompile the program, and test again. A compiled program will usually take much longer to write and test than an interpreted program but, once it is done, it runs much faster.

Compiled programs make use of a collection of ready-made standard machine code routines that are held in *library files*. Some of the routines from the library files are requested by the programmer when writing the source code. When the program is compiled, these routines are added to the compiled program by a program known as a *linker*. The RAM addresses are added and checked by a *loader* program. The final version of the program is then ready to run.

We will now look in more detail at a some of the high-level languages that are useful for programming microelectronic systems.

C compilers

C is one of the most popular languages for programming microprocessors and microcontrollers. Since the C compiler has to be matched to the features of the particular type of CPU it instructs, there are many versions of it. However, the American National Standards Institute has drawn up a set of rules for the way in which C programs are written and most compilers follow these rules. These compilers are known as ANSI C compilers. There is also a version of C with additional features and this is known as C++. There is also a version of C++ called Visual C++. This is a Windows program, primarily intended for writing applications based on a graphical user interface. It is more difficult to learn than a simple C compiler program.

There are rules for typing in a C program and these rules must be strictly followed if the program is to be compiled without error messages appearing. As with an assembler, all variables used in the program must be declared, preferably at the beginning of the program.

Test your knowledge 8.1

What is the function of the symbols /∗ and ∗/ on the first line?

Test your knowledge 8.2

What is meant by 'int' on the third and fourth lines of the program?

Example:

Here is the sample program of Chapter 7, re-written in C:

```
/* Example 1 re-written in C */

int counter
int trial
#define BODY 37

main()

{
counter = BODY;
trial = ++counter;
```

After the initial declarations and a definition, the main program is very short and consists almost entirely of readable (and memorable) words. It is enclosed in braces {} but, as this is only a section of the sample program, the final brace is not shown in the listing above

The program declares two variables, counter and trial as integer variables. Note that we do not have to specify where these variables are to be stored or even whether they are to be stored in a register of the CPU or in a location in RAM. The compiler takes care of all the details and always knows where to find the values of counter and trial when it needs them.

The program also defines a constant, BODY. This will hold its value for the whole program. We have typed its name in capitals to indicate (to people looking at the program) that this is a constant, not a variable.

As in several other languages, calculations are performed by *assigning* a value to a variable. In C, this is done by using the 'equals' symbol. In the first line of the main program, variable counter is assigned the value of the constant BODY. We have already assigned the value 37 to BODY, so BODY holds 37. The first line of the main program then assigns the value of BODY to counter. Then both BODY and counter hold 37.

Perhaps the only item that needs an explanation is the 'double plus' on the last line of the sample. This line means 'increment counter and assign its new value to trial'. The operation could also be programmed as:

```
counter = counter +1;
trial = counter;
```

The equals sign clearly does not mean equals in the usual sense. 'Counter' can never *equal* 'counter + 1'. Instead, it means assign counter the value of-counter + 1

The sample listed above does not have any label indicating the program line that the CPU is to jump back to later. This is because C does not have this feature. There are several other ways of making a program section repeat, but we will leave descriptions of these until later.

Most C compiler programs comprise several sections for handling the different stages of production of the final machine code file. The first section to use is the Editor, which is rather like a word processor. When the program above, or a sufficiently large part of it, has been completely typed in, using the Editor, you click on the 'Compile' button. The compiler will attempt to compile your program, but will report any errors you have made that prevent it from compiling. If there are errors, you return to the Editor to correct them. When all errors have been eliminated, the program compiles the complete listing into machine code and produces a new file with the .C extension. This is an executable file that can then be run, perhaps on the computer or after downloading into the ROM of a microcontroller. It can also be saved on to disk.

Extension: the group of up to three letters at the end of a file name. It is separated from the main name by a stop. Extensions are used to indicate file types. Here .C means a compiled C file.

There is a lot more to writing programs in C than this example illustrates, but it is enough to help you compare C with the other languages described here. There is more on C in Chapter 11.

BASIC

The acronym BASIC stands for Beginners All-purpose Symbolic Instruction Code. It is probably the easiest language to learn, so it is ideal for beginners, but this has led many people to reject its use for 'serious programming'. However, the fact remains that one can do almost anything with BASIC. Moreover, several of the recent prototyping systems use BASIC as their programming language. A notable one is the BASIC Stamp. Its two versions, Stamp 1 and Stamp II, program the PIC16C56 and the PIC16C57 microcontrollers respectively, using a special Stamp version of BASIC.

The BASIC programs in this book are written in GW BASIC. Many people use QBASIC, which has been developed from GW BASIC. Programs written in GW BASIC will work when typed into QBASIC and also TURBO BASIC.

We shall return to Stamp BASIC later, but first we will look at the original BASIC. There have been many versions of BASIC written for different computers. This lack of standardisation contrasts with the strict definition of C by ANSI. This is one reason why BASIC has been less popular with the professional programmers, who like to be able to write a program on one model of computer and quickly adapt it for running on other computers. The original BASICs differ from most other languages in that the program lines are numbered. Line numbers are keyed in as the program is being written. Conventionally, the lines are numbered in tens. This allows extra lines to be inserted in the program later, without having to disturb the earlier numbering. Here is the sample program written in BASIC.

Example:

```
10 REM Example 1 written in BASIC
20 body = 37
30 counter = body
40 counter = counter + 1
50 trial = counter
```

Line 10 begins with REM, which stands for 'Remark'. The CPU ignores everything on the line following the REM statement. The remainder of this program sample is easily read and understood, provided that we remember that the '=' symbol means 'assign the value of the variable, constant, or expression on the right to the variable on the left'.

Setting out a BASIC program is much simpler than a C program. It is not necessary to begin by defining variables and constants, though you can do so if you prefer to. If you do not define variables, BASIC sets aside locations in memory for them the first time each variable is used. There is no use of punctuation marks at the end of program lines in BASIC. Program lines just end with a carriage return (ENTER on the keyboard) instead of a semicolon. There is no use of braces for limiting sections of the program. In most versions of BASIC, line numbers are used instead of labels.

Example:

If, when the CPU gets to line 250, we want it to jump back to line 40 to increment `counter`, we can use this statement at line 250, later in the program:

```
250 GOTO 40
```

Many programmers prefer to avoid using GOTO, as it may lead to sloppy programming that is difficult to understand later. Instead, they would use a properly constructed program loop, as we shall describe later.

The same program written in PBASIC for the Stamp 2, looks different from the program above. Part of the reason is that the microcontroller has a limited amount of RAM (only 2 K) so the program must be as compact as possible. One obvious difference is that there are no line numbers. This means that there must be labels on certain lines:

```
'Example 1 written in PBASIC

counter var byte
trial   var byte
body    CON 37

                counter = body
again:          counter = counter + 1
                trial = counter
```

Test your knowledge 8.4

What does 'CON' mean in the fifth line? (Count the blank lines too.)

The program begins with a title and, in PBASIC, we use an apostrophe to indicate a comment. It is so much easier to understand what is happening in PBASIC programs that it is rarely necessary to append remarks to every line. Note the label `again:` beginning the line to which the CPU is to return later. Later in the program the CPU could be sent back to this line by the statement:

```
GOTO again
```

As in the previous program, it probably would be better not to use GOTO. A properly constructed loop should be used instead.

PBASIC has a useful feature for debugging programs. To find the values of all the variables after the program has been executed, we append these debugging statements to the program:

```
DEBUG DEC body
DEBUG DEC counter
DEBUG DEC trial
```

This instructs the CPU computer to obtain the decimal values of the variables and the constant and send them back to the computer along the programming lead. The computer displays them on the screen in decimal form. As soon as the program has run, a panel appears on the computer monitor, with these numbers on it:

```
37 38 38
```

Debug commands can be inserted anywhere in a program to show the values of variables at any stage.

The Stamp is primarily intended for use in control or measurement systems. It has a number of features that fit it for this, and therefore PBASIC includes a number of commands for using these features.

These include:

BUTTON	debounce a button and jump to a different part of the program.
PULSOUT	output a timed pulse.
DTMF	generate DTMF dialling codes.
POT	read the value of a potentiometer, thermistor or any other sensor with variable resistance.

These and other built-in routines save the programmer much time.

Visual BASIC is one of the relatively new languages that exploit the facilities of the Microsoft Windows environment. It has most of the keywords of conventional BASIC and a great many more of its own. Most of these are aimed at providing the user with the means of setting up the complicated and visually effective windows, menus, radio buttons, sliders and other graphic devices that are associated with Windows programs. For those wanting to write their own Windows utilities and games programs, Visual BASIC is an ideal platform. Unfortunately, its keywords do not include the INP and OUT functions of conventional BASIC. To access the ports of the computer requires additional programs which may be downloaded from the World Wide Web, but there is no guarantee that these will always be available.

Ladder logic

Programmable logic controllers play an important role in industrial and other control systems. Many of them are programmed by keying a *statement list* into a PC (Fig. 1.4). The statement list is much like a program written in BASIC, though it has its own keywords and syntax.

Most PLCs can also be programmed in *ladder logic*. This is not so much a high-level *language,* as a visual way of representing a sequence of control instructions. It was designed to be programmed by electrical engineers rather than by software experts.

Ladder logic is derived from relay logic, which was widely used in industrial control systems. The plant was controlled by a system of hard-wired relays for switching on lamps, motors and other items of equipment. The diagrammatic form of the program is based on the symbols formerly used in USA for relays and associated devices. Actuators are represented by circles.

Fig 8.1 shows part of a PLC 'program'. With the control unit in programming mode, the diagram is entered on the screen. The keyboard of the control unit has special keys for each of the relay logic symbols. If a PC is being used for programming, certain keys are allotted to display the symbols. The program 'lines' are numbered (the example shown is line 46) and the symbols are each identified with a number or code which indicates memory locations corresponding to terminals on the input or output cards. The system used on Mitsubishi

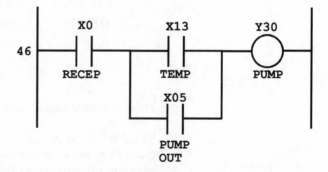

Figure 8.1 *One line of a program for a PLC which controls a pump to pump liquid out of a heating tank. This happens when (1) there is a receptacle ready to receive it and (2) when the liquid has reached the required temperature or the 'pump out' switch is closed manually. Symbols marked 'X' are inputs, and that marked 'Y' is an output. The numbers refer to the I/O cards that are wired to the corresponding sensors and actuator.*

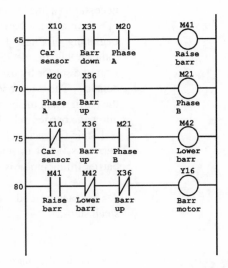

Figure 8.2 *A 'program' that controls the raising and lowering of the barrier of the car park. It has the appearance of a ladder with four rungs, which is why PLC programs are described as ladder logic.*

PLCs uses the 'X' prefix for inputs and the 'Y' prefix for outputs. Annotation on the diagram helps the user to understand how the system functions.

Reading across from left to right in Fig 8.1, we see that, if the receptacle is in place (a proximity sensor detects this and location X1 goes logic high), we can trace a continuous path across its symbol. Then, if the temperature is correct (another continuous path if it is) we can continue across to the symbol representing the pump motor. This complete the path from left to right. 'Current' flows along the path and pumping begins. The diagram shows an alternative path, through a manually operated switch. This means that an operator can start the pump, provided the receptacle is in place, but it is not necessary for the temperature to be correct.

Fig. 8.2 sets out an elementary program for controlling the entrance to a car park. In the first line of the excerpt shown here, if an optical sensor detects that a car is waiting to enter, and the barrier is down (as sensed by a limit switch) and the program is in Phase A (barrier down, ready for car to arrive), a memory location (M41) in the controller is set to a '1'.

The next line is line 70. Note that the lines need not be numbered consecutively. This makes it easy to interpose additional lines if necessary as the program is developed. In line 70, if Phase A is operative and the barrier is up, this marks the end of Phase A and the beginning of Phase B (barrier raised, car drives in). M42 is set to indicate this. In line 75, the first symbol inverts the action of the signal from the sensor by using a the symbol of a relay with normally closed contacts. These are open if the car is there. If the car is *not* waiting at the barrier, and the barrier is up and it is Phase B, then the flag is set to allow the barrier to be lowered. In line 80 if the flag to raise the barrier is 1 (this flag was set or not set in line 65), and the flag to lower the barrier is 0 (set or not set in line 75), and the barrier is not up, the barrier lifting motor is started (a Y output).

As in machine code and assembler, the program proceeds in many small steps.

Problems on high-level languages

1 What are the main differences between programming in assembler and programming in a named high-level language?

2 Explain the difference between a compiler and an interpreter.

3 Describe the main features of a named high-level language.

4 What is ASCII code? Give some examples.

5 Explain what is meant by *assignment*.

6 What is ladder logic and in what applications is it used?

9 | Mnemonics

Summary

Short assembler programs are used to demonstrate how a CPU performs elementary tasks. The mnemonics investigated are those for arithmetic and logical operations, data transfer, branching, and bit handling.

Whatever language is used for programming, the program used by the CPU consists of a series of opcodes and operands. In an assembler program, the opcodes are represented by mnemonics. These may be divided into a number of types, according to what they do. Below we will look more closely at a few of each type of mnemonic. They are demonstrated by using them in short assembler programs. These assembler programs are written for the '1200' microcontroller. Similar operations are available on all microprocessors and microcontrollers, though they may differ in the mnemonics used to identify them.

Arithmetical mnemonics

These cover such operations as addition, subtraction, incrementing, decrementing, and clearing or setting registers. Addition and subtraction can only be done with two numbers. Adding two numbers in binary may produce a '1' to carry over to the next stage. This is

Architecture and programming

CPUs vary in the ways they handle data. Some do almost all the processing in a special register known as the *accumulator*. Some have a few additional registers, which assist the accumulator. Others, for example, the '1200' microcontroller, have no accumulator. Instead, it has 32 general-purpose registers, of which 16 can be used for arithmetical and logical operations.

The instruction set of a given CPU and the corresponding mnemonics of the assembler depend very much on whether processing is centred on the accumulator, or spread over several registers. Whatever the architecture, the general principles of processing in short simple steps is the same for all CPUs. The short programs in this chapter are easily adaptable to a range of CPUs.

stored as the carry digit (C) in the status (flag) register.

ADd with Carry (ADC) is used at the second and subsequent stages of a multi-stage addition. A short program to test this is:

```
;Demonstrating ADC, adding 149 + 47 + 1 (carry)

sec                 ;set carry flag.
ldi r16, 149        ;load r16 with 149.
ldi r17, 47         ;load r17 with 47.
adc r16,r17         ;add all three and put the
                    ;sum in r16.
```

Test your knowledge 9.1

What would the result be if the last line of the pprogram was changed to adc r17,r16?

Addition in stages

CPUs operate on pairs of numbers. They can add *a* and *b* to obtain their *sum*. If there are three numbers *a*, *b*, and *c* to be added, it has to be done in two stages:

a + b = sum1
sum1 + c = sum

There is an exception when two numbers and a carry bit are being added. All three are added in a single operation, as in the ADC operation, described on this page.

To prepare the program for demonstrating ADC, the carry flag is set and the two numbers are loaded. The registers hold this:

Flags	I	T	H	S	V	N	Z	C
	0	0	0	0	0	0	0	1
R16	1	0	0	1	0	1	0	1
R17	0	0	1	0	1	1	1	1

After the final line of the program (ADC), the registers hold this:

Flags	I	T	H	S	V	N	Z	C
	0	0	1	1	0	1	0	0
R16	1	1	0	0	0	1	0	1
R17	0	0	1	0	1	1	1	1

> **Test your knowledge 9.2**
>
> Why is the half-carry flag set at the end of this program?

The addition has not produced a carry (that is, there is no ninth bit) so C has been reset. Some other flags have been set, but these are not relevant to this calculation. Now r16 holds the sum (= 197), and r17 still holds the value with which it was loaded.

SUBtract without carry (SUB) ignores the state of C. A short program to test this is:

```
;Demonstrating SUB, subtracting 47 from 149.

ldi r16, 149        ;load r16 with 149.
ldi r17, 47         ;load r17 with 47.
sub r16,r17         ;subtract r17 from r16 and
                    ;put the result in r16.
```

> **Test your knowledge 9.3**
>
> Whet would be the result of appending `inc r17` to the end of this program?

The carry flag may be '0' or '1', depending on the result of a previous operation. After the numbers have been loaded, the registers hold this:

Flags	I	T	H	S	V	N	Z	C
	0	0	0	0	0	0	0	0/1
R16	1	0	0	1	0	1	0	1
R17	0	0	1	0	1	1	1	1

After the final line of the program, the registers hold this:

Flags	I	T	H	S	V	N	Z	C
	0	0	1	1	0	1	0	0
R16	0	1	1	0	0	1	1	0
R17	0	0	1	0	1	1	1	1

No carry is produced so C has been reset. R16 now holds the difference (= 102), and r17 still holds the value with which it was loaded.

Logical mnemonics

Logical operations are bitwise (between corresponding pairs of bits), as these examples show. There are no carries with logic, so the C flag never gets set.

AND is demonstrated by a short program, using the same two values used in the operations above. After the values are loaded, the registers hold:

Flags	I	T	H	S	V	N	Z	C
	0	0	0	0	0	0	0	0
R16	1	0	0	1	0	1	0	1
R17	0	0	1	0	1	1	1	1

After the final line of the program, the registers hold this:

Flags	I	T	H	S	V	N	Z	C
	0	0	0	0	0	0	0	0
R16	0	0	0	0	0	1	0	1
R17	0	0	1	0	1	1	1	1

Only in the first and third columns from the right is there a '1' in r16 AND r17.

EOR (called XOR in some instruction sets) is the logical operation of exclusive-OR. The result is '1' if either of the two bits (but not *both* of the bits) are '1'.

Before operating on the same two values as before, we have:

Flags	I	T	H	S	V	N	Z	C
	0	0	0	0	0	0	0	0
R16	1	0	0	1	0	1	0	1
R17	0	0	1	0	1	1	1	1

Test your knowledge 9.4

Write an assembler program to demonstrate the action of the AND mnemonic.

The result of EOR is:

Flags	I	T	H	S	V	N	Z	C
	0	0	0	1	1	0	0	0
R16	1	0	1	1	1	0	1	0
R17	0	0	1	0	1	1	1	1

Register r16 now holds the exclusive-OR result, and r17 still holds the value with which it was loaded.

Data transfer

There are instructions for transferring (or copying, to be more correct) data from one register of the CPU to another. There are also instructions for sending output to a port, and for reading in data received by a port. With many CPUs, data can also be copied from a register to an address in RAM, or from an address in RAM to a register.

OUT copies data from a general-purpose register to a port register. A short program demonstrates the action of OUT:

```
; Demonstrating the action of OUT

ldi r19, 255          ;all 1's
out $17, r19          ;to ddrb

ldi r20, 165          ;binary 10100101
out $18, r20          ;to portb register
ldi r20, 0            ;binary 00000000
out $18, r20          ;to port b register
```

Test your knowledge 9.5

How would you set up the port with the top nybble as inputs and the bottom nybble as outputs?

The first step is to configure all lines of Port B as outputs. This is done by using out to write 11111111 into data direction register B, which has the address $17 in the I/O memory space. Then the data from r20 is copied to the Port B register, which has the address $18. This could be checked by looking at the port register using simulator software, or by using a multimeter to measure the output voltage at each pin.

In the '1200', '0' means a logic *high* output, so D1, D3, D4 and D6 should be high, and the rest low. In the last two lines, out is used a second time to make all outputs high. You will need to single-step the program to give time to check outputs after the sixth line.

Test your knowledge 9.6

How would you amend the program to make the MBB output high and the rest low?

ST tells the CPU to STore data in one of the registers. ST is followed by the number of the register in which the data is to be stored. However, unlike ADI, it does not immediately declare the register mov ra, rb copies data from rb to ra.

Since Z always means r30, it might be wondered why it is necessary to tell the CPU that the data is in Z. This is because some other CPUs in the family have three registers X, Y, and Z that can be used with ST.

number. Instead, it gives the CPU the number of *another* register (r30, also known as register Z) which is holding the number. This is rather an indirect way of copying data so it is referred to as *store indirect*. The end result is the same as if we used MOV, but the indirect feature is useful in certain programs, as we shall show later.

A dry run helps explain the action of ST:

Program	Z	R16	R20
ldi r30,20	20	0	0
ldi r16,125	20	125	0
st Z,r16	20	125	125

Test your knowledge 9.7

Adapt the program to load $8B into r19, and then copy it to r17.

Indirection

This is a technique that is often used in programs. Instead of giving the CPU a value directly, the CPU is told the number of a register where the value is already stored.

Example:

ldi r19, 48 is a direct or immediate command. It means 'load register 19 with the value that follows immediately (48)'.

st X, r16 is an indirect command. It means 'look in register X (actually r30), find the register number that is stored there, and then store the contents of r16 in the register that has that number'. Register X acts as a pointer to another register.

Indirection makes programming more complicated but it has a useful feature. A direct address (the '48' in the example above) is written into the program. It cannot be altered while the program is running. With indirection, the number in X may be changed as the program runs. Then it may point to any of the other registers. In this way the value in r16 may be stored in different registers, depending on the stage the program has reached.

The number of the destination register (r20) is placed in the Z register (r30). Then we put some data (125) in r16. ST tells the CPU to copy the data present in r16 into the register pointed at by Z.

Branch instructions

A branch instruction tells the CPU to stop working its way systematically along the program and to jump to some other part of the program instead. The jump may be forward or backward.

RJMP, or Relative JuMP, is the simplest of the branch instructions. Normally the program counter is incremented after each fetch-execute cycle so that the CPU automatically fetches the next byte in the program for execution. RJMP puts a completely new address in the program counter, so the next fetch-execute cycle starts at a different place in the program.

The program below explores the action of RJMP in the forward direction:

```
; Demonstrating the relative jump, RJMP

               ldi r16,10
               rjmp land
               ldi   r16,20
               ldi r17,20
               ldi r18,20
               ldi r19,20
land:          ldi r20,20
               ldi r21,20
               ldi r22,20
```

Test your knowledge 9.8

What would happen if the program was edited to put the label on the sixth line from the end?

The CPU jumps to land on the third line up, so registers r20, r21 and r22 are loaded with '20'. Registers r16, r17, r18, and r19 are left unchanged.

Sometimes we want the CPU to continue working its way along the program *provided that* a given condition is true. However, if the condition is not true, we want it to branch to some other part of the program instead. This is known as a *conditional branch* or jump.

BREQ, or BRanch if EQual, makes the CPU jump to another point in the program if the result of a calculation is zero. Before making the jump, the CPU goes to the status register to find out if Z is set (= 1) or reset (= 0). The zero flag is set to 1 if the result of a previous operation is zero. If Z = 1, the CPU jumps to the branch address.

This program demonstrates the action of BREQ:

```
;Demonstrating BREQ

                ldi r16,5
repeat:         dec r16
                breq done
                rjmp repeat
done:           ldi r17,100
```

The program introduces another arithmetic opcode, DEC. This DECrements a register (reduces it) by 1. Here is the beginning of a dry run of this program. Column Z shows the zero flag, and PC is the program counter (program lines are numbered from 0):

The program counter holds the address of the opcode currently being executed.

Program	R16	R17	Z	PC
Ldi r16,5	5	0	0/1	1
repeat:dec r16	4	0	0	2
breq done	4	0	0	3
rjmp repeat	4	0	0	1
done: ldi r17,100	3	0	0	2

In these dry run tables, the content of registers (including the PC) are shown *after* the line has been executed.

The zero flag may be set or not to begin with, depending on the result of the previous calculation. Register r16 is loaded with '5' and the first decrementing results in '4'. The CPU runs round lines 1, 2 and 3 repeatedly, with r16 being decremented each time round. Eventually the CPU is sent back to repeat at line 1,with r16 holding 1 and with Z still zero:

These programs can be tested by doing a dry run, as shown here, or by assembling them and then testing them with a debugging program.

Program	R16	R17	Z	PC
repeat:dec r16	0	0	1	1
breq done	0	0	1	2
[jump to done, at line 4]	–	–	–	–
done: ldi r17,100	0	100	1	4

This segment begins with the zero flag being set when R16 is eventually decremented to zero.

The routine ends with r16 = 0 and r17 = 100. The PC shows that the CPU has gone to line 4. The zero flag is still set because there have been no subsequent operations that involve it.

Bit and bit-test instructions

These are concerned with handling individual bits in a register, and setting or resetting some of the flags in the status register.

ROL is short for ROtate Left. Each bit in the register is shifted one place to the left. The operation involves the carry bit (C) of the status register. When the bits are shifted, the bit in carry is shifted into the register as the LSB. The MSB in the register is shifted into the carry bit.

Suppose that a register holds these bits:

C				Register				
0	0	1	1	0	1	0	0	1

The register holds 105, in binary. A left-shift puts 0 into the LSB and shifts 0 into the carry:

Test your knowledge 9.9

How would you divide a binary number by 2?

C				Register				
1	1	0	1	0	0	1	0	0

The register now holds 210. Shifting a binary number one place to the left is equivalent to multiplying it by 2. Left shifting is the basis for some multiplying routines. After a second left-shift we have:

C				Register				
0	1	1	0	1	0	0	1	0

The LSB holds 0 from the carry, but the carry now holds 1 from the previous MSB. Including the carry as a ninth bit, the value now equals 420. It has doubled again. A third shift gives:

C				Register				
1	0	1	0	0	1	0	0	1

The LSB holds 1 from the carry bit. The carry holds 1 from the previous MSB. There is no doubling this time because of the rotation. The carry bit has gone into the LSB instead of being shifted along to give a tenth bit.

CLZ, or CLear Zero flag, changes the zero flag from 1 to 0, or leaves it as 0 if it is already 0. It is an example of several similar instructions for altering the carry flag (C), the negative flag (N), and others.

The programs above are intended simply to demonstrate some of the things that a CPU can do. Now we will look at ways in which these actions of the CPU can be used to do something useful. Before a system can process data, it must be able to accept data from the outside world. After it has processed the data, it needs to be able to pass it back to the outside world. The next chapter describe routines for inputting and outputting data.

A loop

The program demonstrating BREQ (p. 140) includes our first example of a loop. The CPU is made to repeat a section of the program. This is a *conditional loop* because the CPU is made to jump out of the loop when a certain condition becomes true. In this example, the condition is that r16 has been decremented to zero. The branch command tests the zero flag (Z) and makes the CPU jump out of the loop when Z = 1.

Activity 9.1 Investigating mnemonics

The short programs in this chapter are based on the instruction set of the '1200' microcontroller, using the Atmel AVR Assembler for Windows. It is likely that the reader will be using a different microcontroller or microprocessor and will be programming it with a different assembler, or directly in machine code, or in a high-level language such as C or Basic.

The programs in this book are intended to illustrate the important features of processors in general and ways of programming. They show the reader how to tackle the programming of any processor using any programming system. The reader should not take these examples as something to be copied exactly. They are intended as models that the reader will adapt to a different processor and, in doing so, obtain a close understanding of their particular programming system.

1 If you are using an assembler, look through the list of mnemonics used by your assembler and find the mnemonics corresponding to those used in our examples. Write short programs to demonstrate them. Use the examples given here as a guide.

2 If you are using machine code or a high-level language, write short programs, based on our examples, to demonstrate the corresponding actions of your processor.

3 Take each of the programs you have written in 1 or 2 above and rewrite it to do something a little different. For example, write programs to add two different numbers, to add three or four numbers (all less than 64), to subtract different numbers (including an example that produces a negative result). Instead of the assembler programs based on `breq`, experiment with the IF ... THEN statement in BASIC, or the IF statement in C.

4 Investigate three more mnemonics, opcodes, or statements in the language you are using. Write short programs to demonstrate their action.

Problems on mnemonics

In some of these questions you are asked to write a short segment of a program. Base your answer on an asssembler and processor with which you are familiar. Work out the program on paper, testing out your program with a dry run. When your answer is complete, test it by using an assembler program and a real or simulated processor.

1 Describe the action of two arithmetic mnemonics in an assembler program with which you are familiar.

2 Describe the addition of two values, 197 and 88 by a named CPU.

3 Show how a CPU calculates the bitwise AND of $B1 and $AB. What happens to the carry flag?

4 Outline the sequence of steps needed to output a byte of data at an 8-bit port. Write an assembler program to do this, based on an assembler and CPU of your choice.

5 The opposite of the BREQ nmemonic is BRNE (branch if not equal to zero). Write a short program to demonstrate its action.

6 Write an assembler program to multiply an integer number by 4. The maximum value that the number will have is 64.

7 Write an assembler program to multiply an integer number by 6. The maximum value that the number will have is 42. [Hint: This program

involves both rotation and addition.]

8 Select two mnemonics other than those described in this chapter and write an assembler program that uses them both. Add full comments to your program.

9 Rewrite any two of the assembler programs above, using BASIC or C. Discuss the differences between the assembler and the high-level language.

10   Input and output

Summary

A single-bit input or output is all that is needed in many control systems. Assembler and BASIC programs for microcontroller systems, microprocessor systems, and PCs, using single-bit input and output, are presented and discussed. Delay routines are introduced. Two routines suitable for microelectronic control systems are described, with programs in assembler and BASIC.

A 1-bit input or a set of 1-bit inputs is all that is needed in many control applications. In Fig 1.3, one of the two proximity sensors (7 and 8) has a high output (logic 1) when the sliding valve is at one end of its track. At other times their outputs are logic 0.

Similarly, an output may need only one bit to perform a vital task. It takes only a 1-bit output to sound a fire alarm. Although we are mainly concerned with control applications in this chapter, it must be realised that there are many instances of 1-bit input and output in measuring, communications and commercial microelectronic systems.

One-bit input

A 1-bit input is either 0, or 1. The simplest way of inputting this is to use a switch or push-button, as in Fig. 2.6. This is connected to one of the lines of an I/0 port, for example to the LSB line (D0). Before an input can be accepted, this line must be configured as an input. If only this line is being used, the remaining lines could be inputs or outputs. However, it is preferable to configure all unused lines as outputs. If they are inputs and unconnected, it can happen that they pick up stray charge and can appear to have an input of '1'.

Another point to consider is that if we run the program and *then* we operate the switch or button, the program runs so fast that it is likely to have ended before the switch is closed. The input routine must *wait* for the switch to be closed.

Here is an assembler input routine:

```
;One-bit input routine

.equ ddrb = $17          ;ddrb address.
.equ portb = $18         ;port B address.
.equ bit0 = 1            ;set a constant.

ldi r20,254              ;binary 11111110.

out ddrb,r20             ;register 0 as
                         ;input, rest as
                         ;outputs.

start:    in r16,portb   ;read port.
          andi r16,bit0  ;mask bit 0.
          brne start     ;if Z=0, back to start.

ldi r17,255              ;if Z=1, continue.
```

andi: AND immediate

We will discuss this program in a little more detail than there is room for in the comments:

- The first stage in this program is to allocate labels to two addresses in the I/O registers. The data *direction* register of port B, is at address $17, and this address is now conveniently labelled ddrb. The actual I/O registers of port B are at $18, which is now labelled portb.

- As well as labelling addresses, the .equ directive also labels constants. The example here, called bit0, has bit 0 high and the rest low.

- With the '1200', a register is set as an output by setting its direction register to 1. If the direction register is reset to 0, the register is an input. Some processors may work the other way round to this. The program begins by making Port B register 0 an input.

- The waiting routine begins with a label for the CPU to jump back to as it cycles round the next three lines. It first reads the port values into register r16. Then the program uses a technique known as *bit masking*. When the port is read, we are interested only in the LSB. If we AND r16 with 1, or more specifically with 00000001, the result will be 0 if the LSB is 0 and 1 if the LSB is 1. The values of the other bits will not appear in the result, as zero ANDed with 0 or with 1 always gives a zero result. The zero flag is set if the result is 0. Then we use BRNE to test the state of the zero flag. If the result is 1 ($Z = 0$), it shows that the switch has not been closed, so the CPU jumps back to the start label and continues from there. It cycles round the three lines, jumping back repeatedly until the switch is closed and the bit changes to 0. Then $Z = 1$ and the program can continue.

- The last line gives the CPU something definite to do when the switch is closed, so that, when we are debugging the program, we can tell when it has reached the end.

Masking

This is a technique for picking out one or more bits of a byte to see whether they are 0 or 1. It uses the logical AND operation.

Example:

The input at a port is 01011001 and we want to know whether bit 4 is 0 or 1. We do this by ANDing it with a byte in which bit 4 is 1 and the rest 0:

Input is	01011001
AND with	00010000
Result	00010000

The result is positive so the zero flag is 0, meaning that bit 4 is 1. This is tested and the program continues accordingly.

If bit 4 is 0, masking produces a zero result:

Input is	01001001
AND with	00010000
Result	00000000

The zero flag is 1, meaning that bit 4 is 0.

Masking can be used to pick out more than one bit at a time.

Using a simulator and single stepping through the program, we can see the CPU repeatedly jumping back to start. The closing of the switch is simulated as follows. Step as far as breq start and edit the I/O register to make Port B equal to 1 (= switch closed). Resume single-stepping. The CPU jumps back to start once more, then jumps out of the loop at breq the next time round.

Programming in BASIC

This program is for running on a PC. It requires a switch or button (Fig 2.6) connected to line D0 of the 4-BIT I/O (pin 1 of the connector, see Fig. 5.14) . Connect also to one of the ground pins. Make sure that the voltage of the power supply is 5 V DC. Assuming the base address of the parallel port is $0378 (see p.95), the program is:

```
10 REM 1-bit input routine
20 OUT &H037A,&H04
30 state = INP(&H037A)
40 IF state = &H04 THEN 20
50 PRINT "Switch closed"
```

GWBASIC indicates hex by prefixing the number with &H.

Lines are numbered in tens, as is conventional. On the PC port, input is more complicated than output. First we have to make all outputs high, then rely on the input data to pull one or more low. Line 20 uses the command:

OUT *address, value*

The address is the address of the 4-bit I/O register. The value written is 00000100 in binary. The top four bits are not stored (see p. 93). Bit D2 is high, to make the output high. Bits 0, 1, and 3 are low because they are inverted and the outputs will be high. This preliminary setting of the bits is required with this register.

Line 20 is the actual input line, using:

state = INP(address)

This assigns the reading at the port to variable state. The following line displays the value of A on the monitor. The value that is obtained depends on whether the switch is open or closed. If the switch is open the value is &H04, and the program loops back to line 20. If the switch is pressed, state = &H06 and "Switch closed" is displayed at line 40.

Programming in PBASIC

This input routine can be tested by connecting a switch to pin P0, as shown in Fig 2.6. The supply voltage must not be more than 5 V DC. It is best to use the Stamp's own regulated supply available at pin 21.

Here is the One-bit input program, written in PBASIC for the Stamp2:

```
' One-bit input routine

input 0                 ' Make P0 an input.

start:
        debug "Running"

if in0 = 0 then start ' Jump to start if 0.

        debug "Done"  ' Print "Done" if 1.
```

The first line configures pin P0 as an input (the push-button switch is connected to that pin). The waiting routine consists only of the `start` label and an `if … then` statement, in which `in0` is the input read from pin P0.

There are two debug statements to monitor the action of the program. In this program, debug is being used to pass messages back to the programmer that certain stages of the program have been reached. Each time the CPU reaches a debug statement, the message is displayed on the screen in the debug panel.

When the program is run, the CPU goes though the loop repeatedly, causing "Running" to be displayed each time. The word "Running" repeatedly scrolls up the debug panel at high speed. When the button is pressed the word "Done" is displayed just once and the program stops.

The debug statements are deleted when the program has been found to operate correctly, leaving a very short and understandable program. Its three lines are equivalent to eight lines of assembler.

Activity 10.1 One-bit input

If you have not already done so, read the introduction to Activity 9.1.

1 Adapt the one-bit input program to the system you are using.

2 Write new versions of the input program to make it do something slightly different. For example, revise the program so that it responds to a switch connected to bit 3 of Port B instead of to bit 0.

3 Write a program that responds only after the switch has been closed and then opened again. [*Hint*: it has to read a '0', then a '1'].

4 Write a program that responds only after the switch has been closed twice.

5 A very simple security system has two input buttons. To gain access, the user must press button 2, *then* button 1. Pressing them in the order 1, 2, or pressing both at the same time is not allowed. The user has only one chance to press them in the right order. Write a program that functions in this way.

6 Expand the previous system to three buttons. Decide on the order of keying and write the program.

7 A security system has three buttons. Write a simple combination lock program that responds only when a selected pair of the buttons are pressed at the same time.

One-bit output

One of the simplest output devices is a lamp switched by a transistor (Fig. 2.7). An even simpler output circuit is a light emitting diode (LED) with a series resistor to limit the flow of current. The circuit of Fig. 10.1a requires about 10 mA, so this is suitable for a microcontroller output pin, which can usually supply up to 20 mA. Microprocessor outputs can not usually supply as much current, so a transistor switch is needed, as in Fig. 10.1b. Assume in the following program that the LED is connected to pin 1 of Port B. The program

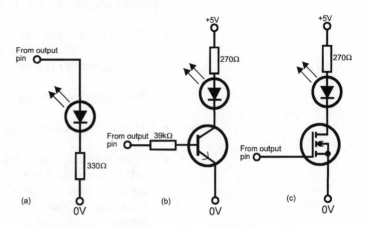

Figure 10.1 *An LED is a useful device for indicating the state of an output port or pin. In (a) the LED is driven directly by the output from a microcontroller. The circuits in (b) and (c) take far less current and are therefore more suitable for outputs taken directly from a microprocessor.*

Test your knowledge 10.1

Why does circuit (c) not need a resistor between the CPU and the transistor?

begins in the same way as the previous one, but then goes on to light a lamp or LED after loading a value into r17.

```
; One-bit input/output routine

.equ ddrb = $17          ;ddrb address.
.equ portb = $18         ;port B address.
.equ bit0 = 1            ;set a constant.

ldi r20,254             ;binary 11111110.

out ddrb, r20           ;register 0 as
                        ;input,rest as
                        ;outputs.

out portb,r20           ;outputs all low.

start:in r16,portb      ;read port.
      andi r16, bit0    ;mask with bit 0.
      brne start        ;if Z=0, back to start.

ldi r17, 252            ;if Z=1. Put binary
                        ;11111100 in r17.

out portb, r17          ;light the LED.
```

Test your knowledge 10.2

What is Z, and when does it go low?

A detailed analysis of the program shows that:

- The program begins the in the same way as before. An extra line before the waiting routine makes all outputs low by storing 1 in each register that is an output.

- The waiting routine is as before.

- The input is *masked* to find the state of bit 0 (see box).

- In the final stage, when the switch is closed, r17 is loaded with a value that has zero at bit 1. A 0 in the output register makes the output go high, so lighting the LED. This is the other way round from many microcontrollers in which 0 = low output and 1 = high output.

Programming in BASIC

The PC has a switch or button connected to bit D0 (Pin 1) of the 4-bit I/O register and an LED connected to bit D0 of the 8-bit register..

```
10 REM 1-bit input/output routine
20 OUT &H037A,&H04      :REM Set up inputs
30 state = INP(&H037A)
40 IF state = &H04 THEN 20
50 OUT &H0378,&H01
```

Lines 20 to 40 are the same as before. Note the reminder to make the outputs high in the 4-bit I/O register, so that pressing the button can pull D0 down. The processor loops around lines 20 to 40 until the button is pressed. Then it drops out of the loop and goes to line 50. There an OUT statement sends a byte to the 8-bit output port, making D0 output high and turning on the LED.

Programming in PBASIC

A push-button or switch is connected to pin P0, as before. An LED is connected to pin P1, as in Fig. 10.1

The BASIC version of the one-bit input/output program is as follows:

```
' One-bit input/output routine

input 0
output 1                'pin P1 is an output
out1 = 0                'LED off

start:

debug "Running"

if in0 = 0 then start

out1 = 1                'make P1 high
```

Two additional lines are needed to switch on the LED connected to pin P1. The second debug statement of the previous program is not needed now.

On running the program the word "Running" is repeatedly scrolled up the debug panel. This stops when the button is pressed, at which point the LED comes on.

The debug line is deleted when the program has been tested.

Activity 10.2 One-bit input/output

If you have not already done so, read the introduction to Activity 9.1.

1 Adapt the example of a one-bit input/output program to the system you are using.

2 Write new versions of the input/output program to make it do something slightly different. For example, revise the program so that it responds to a switch connected to bit 3 of Port B instead of to bit 0. Change the output pin to bit 5.

3 Start with the LED on at the beginning of the run and let closing the switch turn it off.

4 Make the user close the switch (or press a button) twice to light the LED.

5 Program a toggle action. As the switch is repeatedly closed, the LED goes on, then off, indefinitely.

The action of Questions 6 and 7 could be obtained by suitably wiring two switches to one input pin. These would be the hardware solutions. The questions ask for the software solutions.

6 Install two switches on two input pins and program the system so that closing either of these turns the LED on.

7 Install two switches on two input pins and program the system so that the LED lights only when both switches are closed.

8 Install two switches and two LEDs. Write a program to turn one LED on with one of the switches and to turn the other LED on with the other switch.

Delays

Processors work so fast that a short program such as the input/output program seems to run instantaneously. The LED seems to come on as soon as we press the button. The speed of action is advantageous, but sometimes we want to hold the CPU back a little. If the command that switches the LED on is followed immediately by a command to switch it off, the action takes place too quickly for us to see the LED flash. To be able to see the flash, we need to introduce a *delay* between switching it on and switching it off again. In this way, a delay is useful to hold the CPU back for a short time, to let the human operator keep up with it. A delay may sometimes be used to make an action repeat at regular intervals.

One way to produce a pause in a program without actually stopping it is to keep the CPU busy doing something that has no apparent effect. In this program we set it the task of decrementing a number until the number becomes zero:

```
;delay routine

;ldi r16,255        ;the number to work on.

start:
      dec r16        ;decrementing.
      brne start     ;and again,
nop                  ;until Z = 1
```

The branch instruction `brne` is the opposite of `breq`. It means 'branch as long as the number is not equal to zero'. In other words branch as long as Z = 0. To give the CPU something to jump to when it has finished running the loop, we use the mnemonic `nop`, which means 'no operation'. The CPU does nothing at this instruction. Inserting one or more `nop` commands in a program can be used to create very short delays, but only of a few microseconds.

The program above loads r16 with 255 and then decrements it by 1 each time round the loop until it reaches zero. It takes one machine cycle to decrement r16, and two cycles to fetch Z and check whether or not it is 1. This makes 3 cycles for the whole loop, and the CPU runs round the loop 255 times. This takes $3 \times 255 = 765$ cycles. If the system is running at 1 MHz, this takes 765 µs. This delay is too short for operations such as flashing LEDs, but it could be useful when waiting for the output voltage levels of an ADC to settle before sampling its output.

For a shorter delay, load r16 with a smaller number. For a longer delay, we need to load a larger number. Like many other microcontrollers, '1200' has only 8-bit registers. If your microcontroller has 16-bit registers, it can produce delays of up to $65536/3 = 21845$ µs, or 22ms. With only 8-bit registers to work with, we adopt another strategy. We use *two* registers, and program *two* loops, one inside the other. These are known as *nested loops*.

```
;ldi r16,255            ;the first number.

startouter:             ;start of outer loop.

    startinner:         ;start of inner loop.
    ldi r17,255         ;the second number.
    dec r17             ;dec. second number.
brne startinner         ;end of inner loop.

dec r16                 ;dec. first number.
brne startouter         ;end of outer loop.

nop                     ;do nothing more.
```

The loop in which r16 is counted down remains as before, except that its label has been renamed `startouter`. The CPU will run this loop 255 times, as before. The difference is that, within this outer loop, there is an inner loop. It starts at `startinner` and is based on decrementing r17. Each time the CPU runs the outer loop, it comes to the beginning of the inner loop. Then it has to decrement r17 from 255 to 0 before it can continue with the outer loop.

It takes 756 cycles to decrement r17 to zero and it has to do this 255 times. Decrementing r19 takes 3 cycles, so the outer loop takes 759 cycles. Running it 255 times takes $255 \times 759 = 193545$ cycles. At 1 MHz this takes 0.193545, or approximately 0.2 s. This is a reasonable delay for flashing an LED, though some times we may want a delay that is longer.

Test your knowledge 10.3

What is the delay if the clock frequency is increased to 5 MHz?

Test your knowledge 10.4

The starting values of the first and second numbers determine the length of the delay. Which one gives the fine adjustment to the length?

There are several other ways of producing delays. Some microcontrollers have built-in timers for this purpose. Timers are sometimes included on the same chip as another device such as an I/O port. The 68230 PIT is an example of this (see p. 77).

Programming in BASIC

BASIC provides two ways of creating delays. One of these is to make the processor run a loop many times. With interpreted BASIC, which has to interpret the program each time round the loop, quite long delays can be obtained. For example:

```
100 FOR j = 1 TO 1E7:NEXT
110 PRINT "Finished"
```

This takes about 18 s on a PC with a 300 MHz clock. The delay is ten times longer if we increase the '7' to '8', and even longer if we make it larger.

The other technique is to make use of time$. This is a string variable with the format:

$$hh:mm:ss$$

It automatically registers the time since it was reset, in houre, minutes and seconds. At any time in a program, time$ can be set to any chosen value. Conversely, we can read its value at any time. This routine gives a delay of approximately 12 s:

```
100 time$ = "00:00:00"   :REM Resetting
110 IF VAL(RIGHT$(TIME$,2))<12 THEN 110
120 PRINT "Finished"
```

Line 110 evaluates the last two characters in time$ and sends the CPU back to the beginning of the line until the value exceeds 12. Very long delays can be obtained by using time$, and the length of time does not depend on the frequency of the system clock. There are two disadvantages of time$. One is that, being a string variable, it is more difficult to extract the information you need, as demonstrated by line 110 above. The other disadvantage is that resetting time$ in a BASIC program resets the clock in the Windows Date and Time routine. There are ways of using time$ without making this happen.

Programming in PBASIC

Timing in PBASIC has a useful PAUSE instruction. For example:

```
PAUSE 1000
```

This gives a delay of 1000 ms, equal to 1 s. The pause can be up to 65535 ms long, or just over 65 s.

If a longer delay is required, PAUSE may be put in a loop.

```
FOR j = 1 to 60
PAUSE 1000
NEXT
```

This repeats a delay of 1 second sixty times, giving a total delay of approximately 1 minute.

The precision of the PAUSE routine is not high. The ceramic resonator in the timing oscillator has a precision of only ±1% and there are slight delays in reading the interpreted PBASIC.

Programming in PBASIC

A PBASIC version of the delay program uses the PAUSE command to flash an LED once every 0.4 s.

```
'delay
output 1          ;pin P1 as output

again:
out1 = 0          ;LED off
pause 200             ;0.2 s delay
out1 = 1          ;LED on
pause 200             ;0.2 s delay

goto again            ; to repeat the flash.
```

The LED is connected to pin P1. This routine is easily expanded to flash at different rates, and to flash more than one LED.

Activity 10.4 Delays

1 Adapt the delay program to the system you are using.

2 Write new versions of the delay program to make it do something slightly different. For example, revise the program so that it produces delay of 0.1 s or 0.15 s.

3 Use the delay routine to turn an LED on for 0.2s, then off.

4 Write a program to turn an LED on for 0.1 s, then off for 0.2 s, repeating. Try varying the on and off periods.

5 Install four LEDs. Write a program to light them for 0.2 s, one at a time, in order.

6 Install two switches or push buttons. Start flashing an LED when one switch is briefly closed (or a button is pressed) and stop the flashing when the other switch is closed.

7 Install two switches and an LED. Take the two inputs to be coded in binary, equivalent to 00, 01, 10 and 11. Depending on the input, flash an LED 1,2, 3, or 4 times.

8 Install two switches and four LEDs. Depending on the input flash the selected one of the LEDs.

9 Write a delay routine that uses INC instead of DEC.

Temperature control system

The elementary input/output programs in this chapter can easily be adapted for simple control functions. For input and output we need:

- A **sensor** that gives a logic 0 or 1 output, depending on whether the temperature is below or above a set level.
- An **actuator** that is switched on or off by a logical 0 or 1 signal and something appropriate to temperature regulation.

As an example, a cooling system is built up from a temperature sensor, a microcontroller and a motor-driven fan. Fig. 10.2 is the circuit for a temperature sensor, based on a thermistor. The set point of the switching circuit depends on the setting of the variable resistor. If the

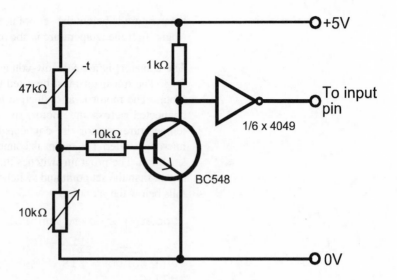

Figure 10.2 *This sensor circuit gives a logic high output when the temperature is above the set point. The supply voltage must not exceed the supply voltage of the processor.*

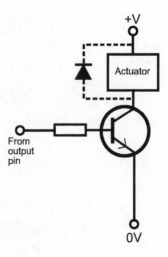

Figure 10.3 *A transistor switch can be used to drive a wide range of DC actuators. The actuator may be an LED, a lamp, a solid-state buzzer, a relay, a solenoid or a low-voltage DC motor. Include the diode when the actuator is inductive.*

temperature is below the set point, the output of the buffer gate is low (logic 0). If the temperature is above the set point, the output is logic 1.

The actuator, is a small low-voltage DC motor, connected as in Fig. 10.3. The transistor must be rated to pass the current required by the motor. The motor is an inductive load so a protective diode must be connected across the motor, as shown by the dashed lines. The temperature sensor is connected to pin 0 of port B of the microcontroller. The motor is connected to pin 1 of Port B and turns a small fan. The program switches the motor on when the temperature is higher than the set point and switches it off again when the temperature falls below the set point.

```
;Cooling system control

.equ ddrb = $17          ;the first 5 lines are
.equ portb = $18         ;the same as in the
.equ bit0 = 1            ;input/output program.

ldi r20,254
out ddrb, r20

sample:

    in r16, portb          ;reading bit B0.
    andi r16, bit0         ;masking.
    breq cold              ;cold or hot?

        ldi r17,252        ;hot, so turn
        out portb, r17     ;fan on.
        rjump sample       ;to sample again.

cold:

        ldi r17, 254       ;cold, so turn
        out portb, r17     ;fan off.
        rjmp sample        ;to sample again.
```

The main points about this program are:

- The initial stages are the same as for the input/output routine, to set up B0 as an input and B1(and the rest of Port B) as an output.
- r20 is not output at the beginning as we do not want to turn on the fan until the temperature has been sampled.
- The sampling routine uses bit masking as in the input/output program.
- If the sample is 0 (= low temperature), the bit is equal zero (Z = 1) and the program jumps on to cold to make the output low. If the sample is 1 (= high temperature) the program runs on to the next line.

- Decimal 252 is binary 11111100, which puts a 0 in the output register. In the '1200', a 0 in the register produces a high output, and turns the fan motor on for cooling.

- In the cold routine, decimal 254 is binary 11111110, which puts a 1 in the output register, turning the motor off.

Programming in BASIC

The sensor is connected to D0 in the 4-bit I/O port (pin 1). The fan motor is connected (through a transistor, see Fig. 10.2) to D0 (pin 2) in the 8-bit output register.

```
10 REM Cooling system control
20 OUT &H0378,0
30 OUT &H037A,&H04
40 state = INP(&H037A)
50 IF state = &H04 THEN OUT &H0378,&H01
ELSE OUT &H0378,&H01
60 GOTO 30
```

Detailed notes:

- Line 20 makes all outputs low at the start of the routine. In a full-length program, this routine could be preceded by a routine which left the register with high levels on some of its lines. Line 20 sets all the outputs to low, ready for this routine.

- Line 30 sets up the 4-bit I/O port in the usual way, with all high outputs.

- Line 40 reads the port and assigns the state of the port to the variable state.

- Line 50 tests state and turns the motor on or off accordingly. If state = &H04, it means that the temperature is high (the equivalent input to 'button not pressed' in the previous programs). The fan is turned on by sending a '1' to the output port. Otherwise, temperature must be low, so a '0' is sent to turn the fan off. The use of IF...THEN...ELSE makes the program shorter and simpler.

- Line 60 sends the CPU back to line 30 to re-initialise the I/O port. If this is not done, the output at D1 stays low indefinitely and the system latches with the fan off.

Programming in PBASIC

The PBASIC version of the cooling system program is as follows:

```
' Cooling system control

input 0          'first two lines are
output 1              'the same as the I/O
                 'program.

sample:

if in0 = 0 then cold

debug "hot "
out1 = 1     'make P1 high to turn
goto sample 'on the fan.

cold:

debug "cold "
out1 = 0                 'make P1 low to turn
goto sample              'off the fan.
```

Security system

To keep the programming simple, this system has been limited to two inputs and one output. The light sensor is a photodiode. A beam from a lamp shines on this. If an intruder comes between the lamp and the photodiode, the beam is broken and a siren is sounded.

There is a reset button, which resets the system, silencing the siren and putting the system on the alert for the next intruder. In real life, it would be best to hide the reset button but in this demonstration system the button is simply connected to pin B2. It is wired as in Fig 2.6 to give an active low output.

The light sensor (Fig .10.4) gives a low output when the beam is broken and is connected to pin B0. The siren is a low-voltage solid-state type, driven by a circuit such as Fig. 10.3 (omitting the diode) and connected to B1.

The listing for the '1200' microcontroller is as follows:

```
;Security routine

.equ ddrb = $17
.equ portb = $18
.equ bit0 = 1              ;Sensor input.
.equ bit2 = 4              ;Button input.

ldi r20,250
out ddrb,r20
out portb, r20            ;Siren off.

waiting: in r16,portb     ;Waiting for intruder.
         andi r16,bit0
         brne waiting

ldi r17,248
out portb,r17             ;Siren on.

alarm:   in r16,portb     ;Siren sounding until
         andi r16,bit2    ;button is pressed.
         brne alarm

ldi r17,250
out portb,r17             ;Siren off.
rjmp waiting
```

Note that after initialising labels and constants (which takes only a few milliseconds) the first action is to turn off the siren. This is to guard against the output going high when power is applied and switching the siren on. This precaution may be unnecessary but it takes care of a possible glitch.

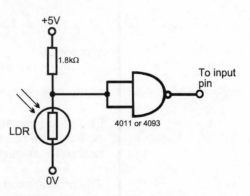

The 4011 is a NAND gate IC. With two inputs tied together, the gate acts as an inverter. The 4093 is the same, except that it has Schmitt trigger inputs, which give a sharper, more definite, on-off action.

Fig 10.4 *The light sensitive circuit has a high output when light is shining on the LDR. It goes low when the LDR is shaded.*

Programming in PBASIC

Connect the light sensor (Fig 10.4) to pin P0 of the Stamp and a siren switched by a transistor to P1. The first version of the security program is similar to the I/O routine on p. 153:

```
' Security system V.1

input 0
output 1
out1 = 0                'ensure LED off.

onalert:

debug "On alert "

if in0 = 0 then onalert

debug "Intruder!"

out1 = 1                'switch on siren
```

If a delay is added, it makes the system insensitive to short breakages of the beam,and so avoids false alarms:

```
' Security system V.2

input 0
output 1
out1 = 0                'ensure LED off.

onalert:

if in0 = 0 then onalert

pause 500
if in0 = 0 then onalert

out1 = 1                'switch on siren
```

The debug statements have been deleted, as they should no longer be necessary. Notice how the beam has to be broken for at least half a second to trigger the siren.

The program can be reset to silence the siren by pressing the restart button on the Stamp, but it may be more convenient to install a pushbutton. Connect a button to pin P2, as in Fig 2.6.

This sequence of versions shows how a program may be

developed stage by stage, testing each stage as it is added. This is not the 'Top down'' approach to program writing so often recommended, but it is helpful to beginners to explore the capabilities of the CPU and the programming language in this way. Here is the third, though not necessarily the final, version of the security program:

```
' Security system V3

input 0
output 1
input 2                    'add input for button

out1 = 0                   'ensure LED off.

onalert:

if in0 = 0 then onalert

pause 500
if in0 = 0 then onalert

out1 = 1                   'switch on siren

alarm:

if in2 = 1 then alarm 'keep siren sounding

out1 = 0                   'siren off
goto start                 'On the alert again
```

Activity 10.5 Projects with 1-bit I/O

The cooling system and security system that have just been described show that it is possible to build useful systems with only one or two sensors and one or two actuators. There are dozens of similar systems that can be built from 1-bit sensors and 1-bit actuators.

1 To begin with, build the cooler or security system, adapting the design and programming to suit the microcontroller, microprocessor or computer that you are using. Build it on a breadboard to start with, though you could transfer it to stripboard or make a PCD for it if it is successful. Try to think of improvements and additional facilities. One practical point: if you are working on the

security system, you should use it to switch on an LED while you are developing it. This is less irritating to other people working in the same room. Later when the design is perfected and the program has been debugged, you can substitute a siren for the LED.

2 Write new versions of the cooler program to make it do something slightly different. For example, revise the program so that it responds to a sensor connected to bit 3 of Port B instead of to bit 0. Change the program to control a motor through an output at bit 7.

3 Adapt the cooler program to switch on a siren when the temperature rises above the set point. This could be a fire alarm system.

4 Adapt the cooler program to flash an LED repeatedly when the temperature falls below the set point. This could be a frost warning system.

5 Substitute other sensors for the thermistor in the circuit of Fig. 10.2. Suitable sensors include a light-dependent resistor, a vibration detector, a pressure mat, and a proximity switch. First test the sensor circuit without connecting it to the processor, and adjust resistance values if necessary. Then decide on a suitable application for the sensor and devise an appropriate actuator. Finally connect a sensor and an actuator to the system and write a program for the application.

6 Design and build a system with up to three 1-bit inputs and up to three 1-bit outputs. For inputs, use circuits sensitive to light, temperature, sound, position or magnetic fields. There are many ways in which switches, keys and buttons of different kinds may be used to provide input. This includes the use of microswitches as limit switches, or to indicate whether doors are open or closed. For outputs use lamps, LEDs, sirens, buzzers, solenoids, electric motors (small ones working on 6V or less, DC), or relays. Aim to control your system with a program not much longer than the programs in this chapter. If it seems that the program needs to be longer, wait until you have studied the next chapter before you write it.

It is intended that you should build the systems on a breadboard to begin with, though some of them will be suitable for transferring to a more permanent form such as stripboard or pcb.

Problems on input and output

1 List three different ways of providing input to a microelectronic system and describe one application of each.

2 List three output devices and describe examples of systems in which they could be used.

3 Which programming language do you prefer to use? Give reasons.

4 What is masking? Give an example of a program in which masking is used.

5 How can we make the CPU wait for a given input, or wait until a particular event has occurred?

6 Why is is sometimes necessary to have a delay in a program? Describe how a delay of 10 s could be produced.

7 Choose *one* of the following tasks and outline the design of a microelectronic system to perform it, using only 1-bit inputs and outputs:

- a radio controlled mechanism for opening and closing a garage door.

- a system to switch on lights at the front of a house whenever someone approaches it, but only at night.

- an automatic coffee vending machine.

- a sliding valve with proximity detectors, as in Fig. 1.3.

- a task of your own choosing.

List the input and output devices that you would use. Outline in words how you would program the system to operate (you are not expected to write the actual program).

8 List the things that should be done at the beginning of a program.

9 Write a delay routine for use in BASIC programs, that uses time$, but does not upset the Windows date and time routine.

10 Basing your answer on the cooler system, design and program a system to control a heater to keep a room at constant temperature.

11 Extend the programming of the security system to switch the siren off after it has been sounding for 15 minutes.

11   Structured programs

Summary

Flowcharts and design structure diagrams are two techniques for displaying the structure of a program. They help the programmer to write efficient programs that will work as intended. Long programs are best broken down into a number of sub-programs, known as modules. Subroutines and functions are modules of a special kind. The operation of the stack and the action of interrupts are two other features of importance to programmers.

The programs described in Chapter 10 are so short that is it easy to read through them and understand what they do. Mostly, they list a sequence of steps that the processor must take, starting at the beginning, continuing to the end and then stopping. Even in some of these simple programs it is necessary for the processor to jump to different parts of the program or to spend some time running round loops. As soon as we introduce jumps and loops into a program, it becomes more difficult for the person reading the program to understand what is going on.

One way of making the action of a program clearer, is to draw a diagram to show the paths that the CPU takes while working its way through the program. A diagram is useful after a program has been written. It is also useful to draw the diagram when the program is being designed and written. Indeed. with a program of more than 10 or so lines, it is essential to draw a diagram *first,* then write the program.

Diagrams of programs

Diagrams help planning, and proper planning is essential if programs are to:

- operate correctly.
- operate efficiently.
- be easily readable and understood by other people.
- be easily readable and understood by the program writer a few weeks later.
- be easily adapted and modified at a later date.

Flowcharts

A flowchart is a method of representing the way a program is constructed. Each small section of the program (perhaps only two or three opcodes or mnemonics long) is represented by a box (Fig. 11.1). Inside the box we write, in a concise way, what happens at that stage. Boxes are connected by arrows, which indicate the direction of flow of the program. The best way to understand the meaning of a flowchart is to look at one drawn to represent one of the simple programs we have already studied.

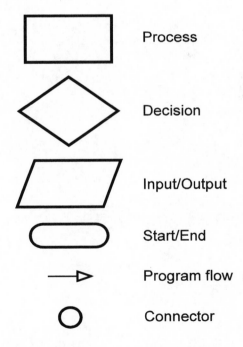

Figure 11.1 *Symbols used for drawing flowcharts.*

Fig. 11.2 illustrates the delay program of Chapter 10. Begin at the terminal box labelled 'Start' and follow the arrows. When you come to a decision box decide whether the answer to the question is 'yes' or 'no' and follow the appropriate arrow. Eventually, after going round the inner loop 65025 times, and round the round the outer loop 255 times, you will finish at the terminal box labelled 'End'. You will find that there is no other route from 'Start' to 'End'. There is no way in

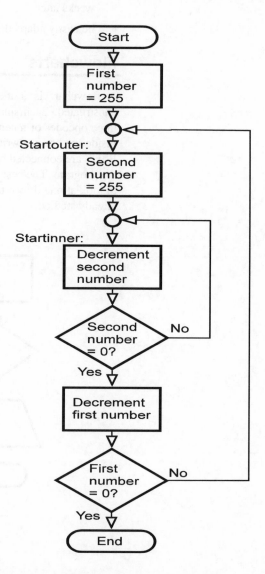

Figure 11.2 *A flowchart of the delay routine clearly shows the nested loops. The line labels are added to the flowchart to make it easier to relate the flowchart to the listed program.*

which the CPU can take a short cut. This proves that the program, if it follows the structure shown in Fig. 11.2, will operate as required.

Fig. 11.3 illustrates the cooler program of Chapter 10. The structure of this is different from the delay program. It has a single loop but the decision box leads the CPU one way or the other, depending on the temperature of the sensor. Note that there is no 'End' terminal in this program. It runs indefinitely (for ever).

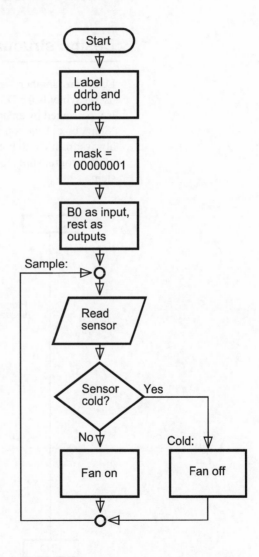

Figure 11.3 *The alternative states of the cooling system are evident in the flowchart.*

Activity 11.1 Flowcharts

1 Draw a flowchart of the security program in Chapter 10.

2 Draw a flowchart of any other program that you have written.

3 Add a third loop to Fig. 11.2. Run it for longer delays.

Design structure diagrams

DSDs are another way of representing programs. Some of the more useful symbols for DSDs are given in Figs 11.4 and 11.5. The program is represented by an unbranched vertical line or 'stem' running from the 'Start' box at the top to the 'End' box at the bottom. Program flow is always from top to bottom, so there is no need for an arrow. The stages of the program are represented by boxes that branch off along the main stem.

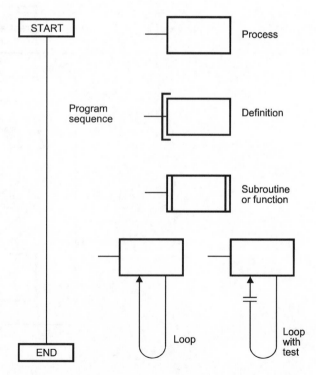

Figure 11.4 *Some of the symbols used in drawing design structure diagrams. The sequence of the program runs from the START to the END.*

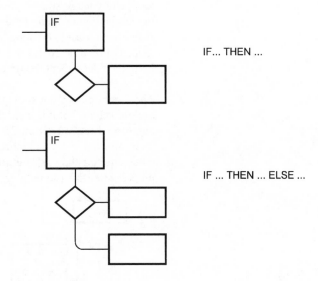

IF... THEN ...

IF ... THEN ... ELSE ...

Figure 11.5 *DSD symbols for processes that depend on a condition being true or not. In the upper diagram, the process occurs only if the condition is true. In the lower diagram a process occurs if the condition is true and a different process occurs if the condition is not true.*

An example makes the DSD system clear. Fig. 11.6 shows the delay routine as a DSD. Instead of following arrows, as in a flowchart, different rules apply. Begin at the top and follow down the 'stem' until you reach a branch. Follow the line that enters the branch until you reach a box. Text in the box describes either a process, a definition, a subroutine or function (more about these later), a loop or a decision. In the figure, the line goes to a box labelled 'FOR first number = 255 to 0'. This is a loop, as indicated by the line beneath it, looping back to the box. The text in the box states how many times you have to run the loop. In this case, you loop until the first number becomes zero.

Like the main stem of the program, the loop line has branches to describe what the CPU has to do as it runs the loop. The first thing is to run another loop, in which the second number is reduced from 255 to 0. This box too has a loop line below it. There is only one branch on this loop. This describes a process, which is 'Decrement the second number'. As you come to this line, enter it, obey the instruction in the box, and leave by the same line. So the CPU runs round the loop line, decrementing the second number each time until it becomes zero. At this point the action of the second loop is completed. So you leave the second loop by the way you entered it. This returns you to the first loop, and you are still on your way round it for the first time. The next branch you enter holds the instruction to decrement the first number. After that, return to the loop box and begin the next time round the first loop. Every time round the first loop you enter and run the second loop, then decrement the first number.

The diagram does not show the nested loop structure in the same way as a flowchart, but it is clear that the CPU has to decrement the second number all the way from 255 to 0, every time it runs round the first loop. When the first number is eventually reduced to 0, the task is complete and the CPU returns to the main stem. There is only one process after this (do nothing) and then the program ends.

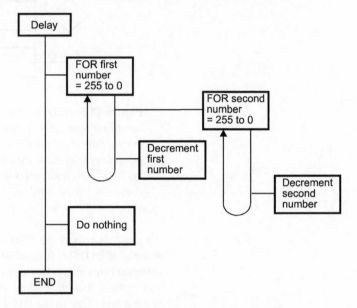

Figure 11.6 *A DSD version of the delay program. The second (inner) loop is reached by way of the first (outer loop), and runs completely for each single run of the outer loop.*

Fig. 11.7 illustrates some other features of DSDs. First there is a sequence of three processes which set up the system ready for action. Then there is a loop, but this is one in which there is no test to see if the task is completed. Consequently, the CPU circles this loop indefinitely. Each time round the loop, it samples the sensor and a decision is taken on whether to switch the fan off or to switch it on. When the decision has been taken and acted upon, the CPU runs back along the way it has come until it reaches the loop. Then it continues around the loop, back to the 'DO' box, and on around the loop again. It samples the temperature and switches the fan accordingly every time round the room. It never reaches the 'End' terminator.

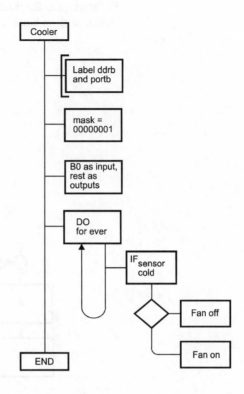

Figure 11.7 *The cooler program, drawn as a DSD, can be seen to run forever, never reaching the end.*

Program structures

The examples we have studied have illustrated the three basic program structures:

- **Sequences**: a series of processes take place one after the other. Figs. 11.2, 11.3 and 11.7 all begin with a sequence of three processes. Some programs consist of nothing but a single sequence, and run straight through, from 'Start to 'End'.
- **Iteration**: the program has a *loop*. The loop may continue indefinitely (Figs. 11.3 and 11.7), or until some condition is true (Figs. 11.2 and 11.6).
- **Selection**: there is a choice between alternative paths, depending on whether a condition is true or not. In some programs only one path (the condition is true) leads to specific action. In other programs (Fig. 11.7) each path leads to an appropriate action. Note that in Fig. 11.7 the selection is within an iteration.

For reference, the three basic structures are shown in Fig. 11.8 as they appear in flowcharts, and in Fig. 11.9 as they appear in DSDs.

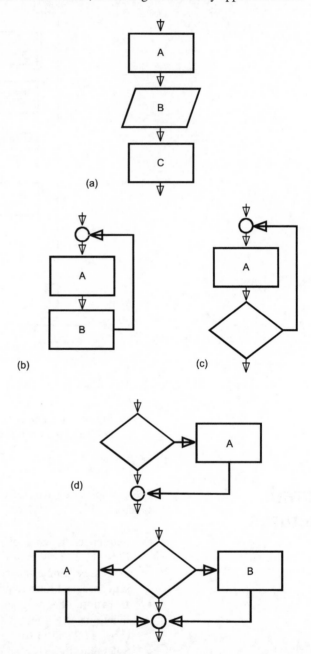

Figure 11.8 *Flowcharts of program structures, (a) sequence of processes A,B and C, (b) iteration of A and B indefinitely, (c) iteration of A until condition is true, (d) selection between performiing process A or not, (e) selection between performing process A or process B.*

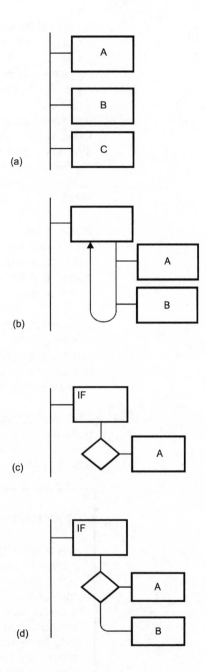

(a)

(b)

(c)

(d)

Figure 11.9 *DSDs of program structures, (a) sequence of processes A,B and C, (b) iteration of A and B indefinitely, or iteration of A and B until condition is true, (c) selection between performing process A or not, (e) selection between performing process A or process B.*

Designing programs

It is generally stated that the best way to design a program is from the top down. The process starts with an outline definition of what the program is to do. With a little experience it becomes clear that most programs can be divided into a number of stages. Usually the first stage, which could be called *initialisation,* is concerned with defining variables, their types, and possibly their initial values. Then come definitions of constants, including masking constants where used. Often certain addresses, such as the addresses of ports and data direction registers, are labelled at this stage, and then the ports are configured as inputs and/or outputs. In some programs it may be advisable to repeat at least part of the initialisation at later stages (see box).

In some high-level languages there is a fixed order in which the definitions must be made. This first part of the program consists of a *sequence* (as defined above) and programming it presents no difficulty except that of making sure that the definitions are correct and complete.

In its simplest form, the remainder of the program falls under three headings: *input, processing, output.* This is certainly true for the control programs used with programmable logic devices, as explained in Chapter 6. It is also true for many other control programs and for measurement programs too.

The idea of breaking down into shorter, simpler sections is not restricted to the whole program. For example, the input section may in turn be broken down into smaller sections, especially if a certain sequence of key-presses has to be made. In this stage the program

Countering corruption

It is a mistake to assume that it is sufficient to initialise control registers (for example, the data direction registers) once and for all at the beginning of a program. Stored data may change, especially in an electrically noisy industrial environment. The change may only be a single bit that changes from 0 to 1 or from 1 to 0, but such a change may have a serious effect on the program. For example, if a port register is set as an input and the data in the direction register becomes corrupted, the line may change from becoming an input to becoming an output. This can have unforeseen effects on the running of the program. It may be be worth while to re-initialise control registers and also essential data such as constants by sending the CPU back to initialising routines at regular intervals.

usually involves iteration while waiting for input. If more than one form of input can be made (for example, when the input device is a keypad or when it is an analogue quantity that is converted by an ADC) the program may use selection to check that the input is valid. It will also use selection to follow different paths, depending on the input values.

The processing stage may be anything from a simple sequence, perhaps only of one or two program lines, to a complex combination of sequences, iterations and selections. Here it is important to use a graphical technique such as DSDs to structure the program. Then the pathways through the program can be checked to confirm that there are no short cuts, and no dead ends. For further confirmation, especially where there are selections, a dry run will help show how the program behaves for different input values. In some cases, the program will remain in the processing stage indefinitely, as in the cooler program.

The output stage is often a simple sequence, because all the work of processing has been done and it remains only to light an LED or switch on a motor. After that, the program may come to an end, or there may be a loop back to input more data, process it and produce new output. However, output is not always simple. The complex actions of a robot are an example of this. Whatever the level of complexity, the main point is to break everything (input, processing, output) down to the level of individual mnemonics or high-level commands. Draw diagrams. Do a dry run. Then program it. It will still need to be tested, which is one of the subjects of the next chapter.

Modular programming

The discussion above assumed that the program is a single input-processing-output chain. Most programs are more complicated than this. For example, the program for a cellphone would have several distinct parts:

- waiting to receive a call
- receiving a call
- handling a call in progress
- terminating a call
- initiating a call
- recovering a number from data store
- dialling
- displaying data
- checking battery level.

Each of these activities can have a self-contained program. Such separate programs within the main program are referred to as *modules*. Each module is programmed individually, perhaps by different programmers. Yet the same basic principles apply to modules that apply to short programs. Each module has input, processing and output. The difference is that the input may take the form of data made available by other modules, and the output may be data required by other modules. The programmer has to ensure that all the necessary data is available, and to know in what variables or in what addresses in RAM it is stored. Conversely, when the module has been run, it must leave the required data in a form suitable for passing to another module, either as the value of a variable, or as a number of bytes in RAM.

Some languages are particularly suitable for modular programming. In C, a self-contained program module is known as a *function*. Using C forces the programmer to think in modular style, and so produce well-structured programs. A function begins with the function name and, in brackets, the names of variables the values of which are being passed to it. Similarly, if a function calls another function, values held in its variables can be passed to a variable in the called function. This is one way of the ways that a function can have input and output. There are other ways of passing values in C, such as by addresses in RAM. Also some function will have input and output through ports in the way we have already studied.

The functions in C are of two kinds. Some are so often needed in so many different programs that they are available in a library file. Examples of these are `printf()`, which prints values and text on the monitor screen. The matter to be printed is included inside the brackets. This is the way the data is being passed to the function. A function of a different kind is `isalpha()`. This is passed a value which may or may not be an alphabetic character. The output of the funtion is 1 (true) if the value is a character, or 0 (false) if it is not. A function such as this is useful when checking input from a keyboard, to make sure that it is of the expected form. There is an extensive range of ready-made functions in C, which make it possible to write programs that are compact and reasonably easy to follow.

The other kind of C function is written by the programmer. An example is a function that is passed the value of a telephone number as input and produces, as output, the signals to drive the dial tone generator. When the functions that a C program requires have either been found to exist already in the standard function library files or have been written by the programmer, they are tied together by the function that all C programs must have, the `main()` function. In length, this may be less than a quarter the length of the whole program. It consist principally of calls to the other functions. These in turn may call the functions that they need. We can see how this structuring of the program mirrors the breaking down of a complex operation into its smallest parts.

Functions in most versions of BASIC are more limited than those in C. BASIC provides a few useful functions, such as SIN, COS and TAN, but the user is able to define other functions as required.

Example:

A measurement system is reading data from an ADC. To convert the data (adc) into a reading of temperature it has to have 0.5 subtracted from it, is then divided by 5.2 and finally have the room temperature (room) added to it:

$$temp = (adc - 0.5)/5.2 + room$$

Data is read many times during the program but, instead of having to type in this equation after every reading, it can be defined as a function, using:

```
DEF FN temp = (adc - 0.5)/5.2 + room
```

This is included early in the program, preferably in the initialisation stage. After that, every time the equation has to be used all that needs be typed is:

```
FN temp (adc,room)
```

It could be used in a statement such as:

```
300 IF FN temp (adc,room) > 29 THEN 460
```

Unfortunately, most versions of BASIC limit function to one line, though some versions allow multi-line definitions.

Subroutines

There are programs in which a particular sequence of operations has to be repeated at several different points in the program. It may, for example, be a routine to set a data direction register of a port to all inputs, read the input data AND with a mask, then store the result in a given register. Much time and memory space can be saved by entering this routine just once, as a subroutine. It can then be called from any point in the program as many times as it is needed. The box explains what happens.

On p. 183 is an example of using a subroutine in an assembler

Subroutines

This illustrates the path of the microprocessor as it runs through a BASIC program that calls a subroutine three times.

```
10 REM Demonstrating subroutines.
   •      (First part of program)
   •
   •
150 GOSUB 600 :REM goes to subroutine
160    (Comes back to second part of program)
   •
   •
210 GOSUB 600 :REM goes to subroutine again
220    (Comes back to third part of program)
   •
   •
   •
350 GOSUB 600 :REM goes to subroutine again
360     (Comes back to fourth part of program)
   •
   •
   •
590 END :REM Stops CPU running on
600     (Start of subroutine)
   •
   •
680     (End of subroutine)
690 RET  :REM returns to line after
```

Each time the CPU finds GOSUB 600, it jumps to that line, executes the program from 600 to 680, then at line 690 jumps back to the line after the one it came from.

program. The reader will recognise lines and routines from the LED switching programs and the delay program of Chapter 10.. The point of interest in this listing is the main program. These are the 5 lines following the flash: label.

These few lines clearly show the sequence of the program. The LED is repeatedly turned on, then off, with a delay in each state so that the flashes are visible. Note the essential mnemonic ret (return from subroutine) at the end of the subroutine. This is rts in some assemblers.

```
; Demonstrating .ORG and subroutines

;Initialisation

.equ ddrb = $17
.equ portb = $18
.equ bit0 = $1
.def ledoff = r18
.def ledon = r19

.org $000              ;program starts $000.

reset:    rjmp start      ;reset routine.
          .org = $004     ;program starts $004.

start:    ldi ledoff, $2
          ldi ledon, $0
          ldi r20, $FE
          out ddrb, r20

wait:     in r16, portb ;input button.
          andi r16,bit0  ;button pressed.
          breq wait

flash:    out portb, ledon    ;LED on.
          rcall delay         ;to subroutine.
          out portb, ledoff   ;LED off.
          rcall delay         ;to subroutine.
          rjmp flash          ;repeat for ever.

; Delay subroutine: same as delay program.

delay:

ldi r16, $FF

startouter:

     ldi r17,$FF
     startinnner:
     dec r17
     brne startinner

dec r16
brne startouter

ret              ;return to main program.
```

Note that we are using hexadecimal for all variable values from now on.

When trying out this program (or your version of it) on a simulator, it reduces run time if a small value such as $5 is substituted for $FF in the delay subroutine.

The reason that the flash sequence is so easily understood is that the reader is not confused by having to read twice through the delay routine. The five steps of the routine are clearly set out, with a line for each. The delay routine has been placed at the end of the listing as a subroutine. Whenever a delay occurs, the CPU is told simply to call the

---

.org directive

A directive tells the assembler how to assemble the program. It is not part of the program. The `.org` directive tells the assembler where in memory to start storing the assembled program. Up to now, we have not used this directive but it is important because, in the '1200', the first 4 bytes of program memory are allocated for special purposes. If the program is stored from byte $000 onward, there is a risk that, when the program is run, its first four bytes may become overwritten with data. The program would be corrupted and would not run correctly.

The first byte of the four is reserved for a jump instruction, telling the CPU where the first byte of the actual program is stored. When the assembler gets to the line `.org = $000` it starts storing at $000. There it stores the opcodes for `rjump start`. After that, it is told to store the program from $004 onwards, so it leaves $001, $002 and $003 empty and begins storing the remainder of the program from $004 onwards.

The effect of this is that, if ever the CPU is reset, its program counter is automatically cleared to $000. It goes to $000 and there finds the instruction telling it to jump to $004. This jump takes it to the beginning of the program proper. From then on, it follows the prgram in the usual way.

---

subroutine, that is, to jump to the address where the subroutine begins. The delay is called twice in this program. By making it a subroutine, we need to include it only once. This saves both typing and storage space in memory. This may be limited in amount in a microcontroller.

Note that in this program we do not pass any data to the subroutine or receive any data from it. This could have been done had it been necesssary, using variables, registers or RAM addresses.

The stack

The stack is a region of RAM set aside for the temporary storage of data. It works in a 'last-in-first-out' manner, like the plate dispenser sometimes seen in canteens. The stack pointer in the CPU holds the address of the address next above the top of the stack.

In some microcontrollers, such as the '1200', the stack is small and is automatically operated. When the CPU goes to a subroutine, the address of the program counter is stored in the stack. This is how the CPU 'remembers' which address in the main program it jumped from. When the CPU comes to the return command (ret) at the end of the subroutine, it goes to the stack, recovers the address and replaces it in its program counter. The value is then incremented by 1 and the CPU goes to the program line following the one from which it jumped.

In other microcontrollers and in most microprocessors, there are opcodes for using the stack. Commonly used mnemonics are psh (= push) to place a given value on the top of the stack, and pop to load a value stored at the top of the stack. Usually the push and pop operations take place between the top of stack and the accumulator register.

As an example of the function of the stack , this table shows a section of RAM and the hex data stored there:

Address	Stored data
3B2C	0
3B2B	0
3B2A	48
3B29	0E
Stack pointer	3B2B
Accumulator	65

The stack pointer holds 3B2B
The accumulator holds 65

The shaded cells indicate the extent of the stack.

The top of the stack is currently at $3B2A but the stack pointer points to the address after this, for this is where the next item of data will be stored. The most recently stored data is 48. Now the PSH opcode occurs in the program and is executed. The data stored is:

Address	Stored data
3B2C	0
3B2B	65
3B2A	48
3B29	0E
Stack pointer	3B2C
Accumulator	65

The stack pointer holds 3B2C
The accumulator still holds 65

The microprocessor may now be engaged in other operations not involving the stack, which remains unchanged. Eventually, with (say)

$B7 in the accumulator:

Address	Stored data
3B2C	0
3B2B	65
3B2A	48
3B29	0E
Stack pointer	3B2A
Accumulator	B7

The stack pointer holds 3B2C
The accumulator holds B7

Now the POP opcode occurs in the program and is executed:

Address	Stored data
3B2C	0
3B2B	65
3B2A	48
3B29	0E
Stack pointer	3B2B
Accumulator	65

The stack pointer holds 3B2A
The accumulator holds 65 again

The stack pointer has been decremented to point to the address one above the new top of the stack. Note that the value 65 is still in 3B2B, but this is now above the top of the stack. The 65 is no longer needed and will be overwritten the next time PSH occurs.

The stack is often used for temporarily storing an intermediate result in a calculation, particularly in processors that have only a few registers. With a simple PSH, a value is quickly pushed from the accumulator on to the stack and may be recovered at a later stage with a simple POP instruction. The stack is also used for registering the state of the processor when a *jump to subroutine* instruction is executed, or when the microprocessor is responding to an interrupt. Before going to the subroutine or interrupt routine, the controller stores the current contents of the accumulator, the status register and the program counter on the stack. As soon as the microprocessor returns from the subroutine or interrupt, it recovers all this essential information from the stack. It then continues operating at the same point in the program that it had reached before it jumped or was interrupted.

Activity 11.2 Subroutines

1 Rewrite the assembler subroutine program in another assembler or a high level language.

2 If you are working in assembler with a simulator program, single-step the program and watch the program counter as it goes to the subroutine. Watch the stack pointer. Try to find the location of the stack in your system. Watch the program counter as the CPU returns from the subroutine.

3 Revise the program so that a value `speed` is passed to the subroutine to control the length of the delay.

4 Revise the flash routine so that the LED flashes faster on each successive loop.

5 Find out about the operation of the stack in the processor that you are using. Write simple programs to demonstrate its action.

Interrupts

The CPU is always busy handling the program, proceeding from one instruction to another, perhaps cycling indefinitely round a loop. Events sometimes occur which require that the processor should interrupt what it is doing and take some other action. Such events include:

- **Arithmetic error:** a typical example is 'division by zero'. A divisor in an expression may on occasion evaluate to zero. Division by zero is not possible mathematically. This kind of error is detected inside the CPU and causes it to interrupt itself. Instead of going on to the next instruction in the program, it jumps to a special *interrupt service routine* (ISR). This might cause an error message to be displayed: 'Division by zero error'. The program is then halted and the programmer checks to find why such an error occurred and alters the program so as to prevent it.

- **Exception error:** Occasionally a CPU may be given instructions that it is impossible to obey, such as storing a batch of data in an area of memory that does not exist. An internal interrupt occurs. The ISR may instruct the CPU to display an 'Exception error' message and then close down the program.

- **Clocked interrupts:** If the system includes a real-time clock, this can be programmed to interrupt the processor at regular intervals. In a data logger, for example, the clock could be programmed to interrupt every minute. On receiving the interrupt, the CPU jumps to the ISR, which instructs it to sample and record input data. In between interrupts is is occupied with general routines such as updating the display and accepting commands from the keypad.

- **Peripherals:** The keyboard buffer may be holding key-presses which the CPU needs to know about, so the buffer sends an interrupt signal (usually a low voltage) level to the interrupt input pin of the CPU. The ISR of the CPU then enables the buffer, allowing it to place the keypresses on the data bus one at a time. The result is further action by the CPU, perhaps storing the ASCII codes in memory, or displaying characters on the monitor screen. A printer is another device that frequently interrupts to tell the microprocessor that it is waiting to receive data for printing.

It can be seen that some interrupts are the result of programming or operating errors, but others are a useful way of regulating the system.

When a CPU receives an interrupt, it:

- Finishes its current instruction first.

- Pushes the contents of its registers on the stack. The content of the program counter is important, so that it can return to the same place in the program after the interrupt routine has been completed.

- Jumps to a fixed address in memory, where it finds another jump instruction giving the address of the start of the ISR.

- Jumps to the address of the ISR and executes it.

- Pops the return data from the stack.

- Resumes processing as it was before the interrupt.

Interrupts by different devices are dealt with according to their priority, because the microprocessor can not handle two interrupts at the same time. An interrupt from the keyboard, for example, may have a higher priority than one from the printer. If the printer interrupts while the microprocessor is dealing with an interrupt from the keyboard, the printer is ignored until the keyboard has been dealt with.

This raises the problem of how the CPU finds out which device is interrupting. With some CPUs there are several interrupt input pins, ranked in order of priority. Devices are connected to these pins according to their priority. Priority is fixed by the wiring. Other CPUs may have only one interrupt line, which is part of the control bus. The CPU knows there is an interrupt but has to find out its source. In some systems there is a line in the control bus called *interrupt query*. When the CPU puts a low signal on this line, the signal automatically goes to all the devices that might possibly be interrupting. The one that is interrupting puts its own device code on the data bus so that the CPU knows what action to take. Another though slower way of obtaining the same result, is for the CPU to interrogate each device in order of priority, reading a flag register in each to find which has its interrupt flag set. Other methods are described later with reference to the Z80.

As an alternative to interrupts, a system may rely on *polling* the individual devices. Each device has a register in which there is an interrupt flag. In between its main activities, the CPU interrogates each device to find out which one or more of them has its interrupt flag set. It then takes action. This system is the slowest of all, as a device has to wait to be interrogated and by then it may be too late to avoid a program crash or a loss of data.

It may happen that the CPU is in the middle of a complicated routine which might fail if interrupted. With some CPUs there is a disable interrupts opcode which causes it to ignore all interrupts, until interrupts are enabled again by using another opcode. We use these instructions to *mask* (shut out) all interruptions while a difficult routine is in progress. This is a matter for the programmer to decide. However, there is usually a highest-priority interrupt that is non-maskable.

The Z80 has two interrupt input terminals. When a logic low level is sent to the NMI terminal, it causes a non-maskable interrupt. The falling edge of the interrupt pulse latches the input, so that there can be no further NMIs until the current one has been dealt with. As soon as it receives the NMI, the Z80 goes to the address $0066. There it finds the address of the non-maskable interrupt routine. Since a non-maskable interrupt is *always* attended to immediately, there is no need for the CPU to acknowledge that it has been received. No acknowledging signal is sent to the interrupting device.

The NMI is generally reserved for an event which threatens to crash the system. For instance an NMI is generated if the power supply fails. On receiving this signal, there is a brief time while the capacitors of the power supply contain enough charge to keep the CPU running. In that short time, essential data in RAM can be saved to disk. When programming a non-maskable ISR for the Z80, it should always end with a special RETN instruction. This not only returns the CPU to the program but also resets the latch on the NMI input.

The maskable interrupt pin of the Z80 ($\overline{\text{INT}}$) has three separate modes of operation, which are set by software instructions. These take second priority to the NMI, so it is essential for the CPU to acknowledge receiving the INT signal. Otherwise, the interrupting device may have ceased sending its signal by the time a current NMI signal has been dealt with and the INT signal will be missed. The Z80 responds to an INT signal by making its $\overline{\text{IORQ}}$ and $\overline{\text{M1}}$ control outputs low.

The Z80 has three types of maskable interrupt:

Vectored interrupts: If the Z80 has been programmed to operate in Mode 0, it acknowledges the interrupt as described above and this causes the interrupting device to put a *restart instruction* or *vector* on the data bus. The Z80 reads this single-byte instruction, which tells it the address of the appropriate ISR. There are eight different restart instructions, sending the CPU to a particular one of eight addresses: $0000, $0008, and so on to $0038. These 8-byte blocks of memory can be used to service the routine or send the CPU on to another address where a longer routine is stored. Having these eight address means that up to eight different ISRs can be programmed and the interrupting device can indicate which one of these should be used.

Direct interrupts: If the Z80 is in Mode 1, it is sent automatically to address $0038, without the need for a restart instruction.

Indirect interrupts: In Mode 2, the Z80 receives a vector as in Mode 0 and adds to this a 16-bit value that is already stored in its I register. By setting different values in the I register, this allows the Z80 to be directed to different blocks of memory, each holding eight restart instructions. This gives the Z80 the flexibility to process a large number of different ISR routines.

Several peripheral devices can be connected to the INT pin, leaving the CPU the problem of finding out which one is interupting. If two devices interrupt at the same time the CPU has to give one of them priority. It is not necessary for them to interrupt at *exactly* the same time. The CPU does not check for interrupts until the end of its current execute cycle so, if two interrupts occur during a cycle, they will both be pending when the cycle ends. One solution to this problem is by *polling*. Either the CPU can interrogate each device in order of priority or there is a register that is attached to the data bus and wired as an input port. When a device interrupts, it sets a bit in this register. The priority of each device can be established by the order of bits in the register. On being interrupted, the CPU reads this register, finds out which devices are interrupting and attends to the interrupt of higher priority.

A second approach to the problem of multiple interrupts is *daisy chaining*. An example of this is seen with the Z80 PIO and SIO (Figs. 5.4 and 5.10) These have an interrupt enable input (IEI) and an interrupt enable output (IEO). The device can interrupt only if its IEI input is at logic high. In the figures, we show IEI connected to the 5 V line, so the chip is enabled for interrupts. The EIO terminal of the device may be left unconnected if there is only one device in the system. However, if there is a second device, the EIO is connected to the EIE of the second device, which has lower priority. Both devices have their INT output connected to the INT input pin of the CPU. The EIO output is normally high but goes low when the first device is interrupting. The low input to the IEI of the second device disables it. It can not interrupt until the first device has finished. The low EIE input also makes the EIO go low. The chain can be extended to three or more devices, so that if the second device, for example, is interrupting, the third and subsequent devices can not interrupt. However, in this case the first device can still interrupt and take priority over the others.

This system of connecting the EIO to the EIE of the device of next lower priority creates a daisy chain in which only the device of highest priority can interrupt when more than one is attempting to interrupt.

Activity 11.3 Interrupts

Find out the interrupt facilties of your CPU. Devise a simple program to demonstrate its action. For example, run a program to flash an LED continually. When interrupted the CPU is to sound a siren once, then return to flashing.

If there is more than one level of interrupts, devise a test program to demonstrate this too.

Problems on structured programs

1 What are the advantages of structured programming?

2 Describe two ways of representing a program as a diagram. Give short examples.

3 Draw a flowchart or DSD of a program that you have written. Add notes to point out structural features such as sequences, iterations and selections.

4 What are nested loops? Give an example to illustrate this structure.

5 What is initialisation? List the actions that the CPU performs in a typical initialisation sequence. Why is it advisable to repeat some of the initialisation steps later in a long program?

6 Explain what is meant by the input, processing, and output stages of a program or routine?

7 What is the top down approach to programming? Give an example.

8 What is modular programming? Give an example of a program that would best be written in modular form.

9 What is a function? Give examples in a high-level language with which you are familiar.

10 Describe the sequence of events when a program calls a subroutine and executes it. What are the advantages of using subroutines in the structure of a program?

11 Describe the operation of the stack. In what ways is the stack used?

12 Under what circumstance may an interrupt be generated?

13 Explain the sequence of events when an interrupt is generated, basing your description on a named CPU.

14 Explain how different interrupting devices may be allocated different levels of priority.

Programming projects

Summary

A detailed study shows how a typical support device may be programmed to demonstrate its facilities. The remainder of the chapter comprises a collection of programming problems, based on the techniques described in earlier chapters, but also introducing some new programming methods.

Programming a support device

Real time clock: a device that can be set to indicate the time, in seconds, minutes and hours (also days, months and years) in a particular time zone.

When designing and building a microelectronic system, it may sometimes be necessary to incorporate one or more complex ICs into the system. The system may need a parallel input port, or a serial output port, or it may need an analogue-to-digital converter. Each of these devices has its own special features and it is unlikely to work unless it is set up in exactly the right way. There is not enough space in this book to describe how to set up and use the many types of I/O ports and other microprocessor support devices. The details are available in the manufacturers' data sheets, which may be obtained from suppliers of components, or often downloaded from the World Wide Web.

This case study describes the initial stages in the design and programming of a system based on a real time clock IC. There are many different real time clock ICs available. The type chosen is the Hitachi 146818. This is a 24-pin IC, intended to be part of a system based one of the 6800 family of microprocessors. It is also suitable for use with many other microprocessors and microcontrollers. Having chosen this IC, it is investigated by following the stages described below.

Data sheet

Data sheets usually begin with a list of the special features of the device. The data sheet of the 146818 explains that this is a time-of-day clock and calendar. It counts seconds, minutes, and hours. It registers days of the week, the date, the month and the year. It can deal with months of variable length and with leap years. It can also be used to generate interrupts at regular timed intervals, and can be programmed to produce a square wave of a number of different frequencies.

The 146818 is a real time clock/calendar that is intended for use in a microprocessor system. It is driven by a crystal oscillator. The output from the oscillator goes to a 20-stage divider to produce a frequency suitable for driving the timing circuits. As might be expected, the logic of the IC is complicated and we shall not describe it here.

The data sheet includes tables of Electrical Characteristics. The most important of these is the supply voltage, which in this case is 5V ± 0.5 V. It is essential to check on this before beginning a project because it may sometimes happen that the operating voltage is not compatible with that of other devices in the system. The Stamp system includes a 5 V regulator, so this can be used to power the clock. It is intended to operate it at 32.768 kHz, and the table in the data sheet shows that the current required is only 500 μA, which is well within the resources of the system.

A data sheet provides a vast amount of detailed information. Coming direct from the manufacturer it is nearly always complete and accurate. However, it is not necessary to read every page of the data sheet. The information that is needed for a given project is often scattered in different parts of the sheet. Although the presentation of information is systematic, it is not necessarily in the order in which it is needed for developing a particular project.The experienced data sheet reader soon learns to skim quickly through the data sheet to ascertain what topics it covers. Then the reader moves rapidly from one part of the sheet to another, picking up facts here and there, and gradually summarising the more relevant ones into two or three pages of notes.

The following paragraphs outline the information which was gleaned from the data sheet and is needed for programming the 146818 as a clock.

Pinout

The pinout (Fig. 12.1) has been marked to show which are inputs and which are outputs. This is done because input pins must nearly always be connected to something. Output pins can usually be left unconnected, at least during the early stages of development. There are eight address *input* pins, which are also data *output* pins. These and four of the control pins are to be connected to a Stamp2, which will be used to command the IC. The data sheet explains that during read and write cycles, the address is placed on the bus by the controller. At a later stage in the cycle, data for writing is placed on the bus by the controller, or data for reading is placed on the bus by the clock.

The data sheet has diagrams to show what other connections need to be made to the IC. The most important is the external clock circuitry, consisting of a crystal, and a few resistors and capacitors (Fig. 12.2). There is a choice of frequencies. In this project we use a 32.768 kHz crystal.

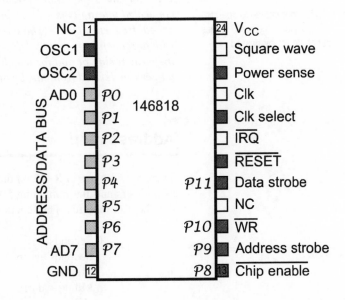

Figure 12.1 *The pinout of the real time clock IC has been marked to indicate which pins are inputs (dark grey) and which are both input and output (light grey). The pins are also labelled with the names of the pins they are going to be connected to on the Stamp2.*

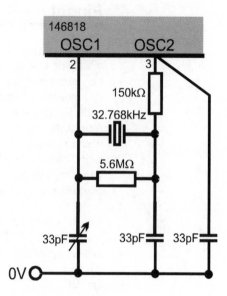

Figure 12.2 *With an INVERT gate connected internally between pins 2 and 3 of the 146818, this circuit oscillates at 32.768 kHz. Also connected internally is a 15-stage binary divider. This divides the crystal frequency by 2^{15}, which is 32768. The result is a squarewave with a frequency of exactly 1 Hz, giving a period of 1 s. This provides the basic timing period for the clock. Periods of 1 min, 1 h, and longer are derived from this by further division.*

When the IC is programmed to operate on a 32.768 kHz clock, the first 5 stages of its divider chain are by-passed.

Address map

There are 64 bytes of RAM on the chip. The first 14 bytes are reserved as registers and the remaining 50 bytes are available for any purpose that the user requires. The functions of the first 14 bytes are:

0 Seconds
1 Seconds alarm
2 Minutes
3 Minutes alarm
4 Hours
5 Hours alarm
6 Day of the week
7 Day of the month
8 Month
9 Year
10-13 Registers A to D

The first 10 registers are loaded with the current times, day, and dates. When the clock runs, a series of divider/counters update the registers as time passes. At any future instant, the registers are read to provide data on time and date.

Registers A to D hold flags to indicate various states of the clock. Some of these flags may be set to control the operation of the clock. For example, it may be set to run in the 12-hour mode or the 24-hour mode.

Data sheets of computer chips usually include a number of timing diagrams, like those in Figs. 7.1 and 7.2. Exact timing is important if the clock is to operate at maximum speed. The diagrams show the minimum times taken for addresses and data to settle on the bus, and the minimum response time of the clock. They show how long data and addresses must be left on the bus to ensure that they are loaded by the clock chip. In general, an IC will work just as well when it is taken through its stages of operation at a slow pace. This project used a relatively slow controller, programmable in BASIC, so there was no need to aim for exact timings. The most important point is the *order* in which control, address and data signals are sent to the clock.

Fig. 12.3 shows a write cycle. The levels on the DS, WR and AS pins have no effect until CE goes low, enabling the chip. The address must now be on the bus and is latched into the clock when AS goes low. Making WR low (= read) with DS high causes the clock to put data on the bus. At the end of the cycle the lines return to their original state with CE high and the rest low.

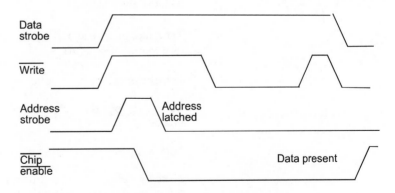

Figure 12.3 *The sequence of voltage levels (high or low) present on the control bus when the CPU is writing to an address in the RAM of the real time clock.*

The next step is to write a program to put a sequence of logic levels on the four control pins:

```
'programming the rtc - writing

loc var byte    'declare variables
value var byte

loc = 2         'their values
value = 12

dirc = 15       'as outputs to control
outc = 1        'disable chip

'write routine

outc = 13       'DS and WR high
outc = 15       'AS high
outc = 14       'enable chip
dirl = 255          'as outputs to bus
outl = loc          'address on bus
outc = 12       'AS low
outc = 8        'WR low
outl = value    'data on bus
outc = 12       'WR high
outc = 1        'disable chip
```

Work through this program to see how the sequence of outc instructions corresponds to the levels shown in Fig. 12.2 (see box).
The values of location and value are edited into the program before the it is run. It would be an improvement to write a proper routine for inputting this data. The program loads the selected location with the given value.

The data can be read back by using this program, which is based on the read routine (Fig. 12.4):

```
'programming the rtc - reading

loc var byte    'declare variables
value var byte

loc = 2         'their values

dirc = 15       'as outputs to control
outc = 1        'disable chip

'read routine

outc = 13       'DS and WR high
outc = 15       'AS high
outc = 14       'enable chip
```

Stamp I/O

The Stamp has 16 I/O pins (P0 to P15), divided into a high byte and a low byte. The low byte (P0 to P7) is accessed by using the label `l`. To set the data direction registers of the low byte, we use `dirl`. To output a value we use `outl` and to input a value we use `inl`.

The lower nybble of the upper byte (P8 to P11) is labelled `c`. To access and use these pins as a block we use `dirc`, `outc` (and `inc`, but not in this program).

The pins of the control nybble are:

P11 (MSB)	P10	P9	P8 (LSB)
DS	WR	AS	CE
1	1	0	1

The bottom row shows the `value` to allocate to the nybble when AS is low and the other pins are high. When programming in decimal, this is expressed as:

$$outc = 13.$$

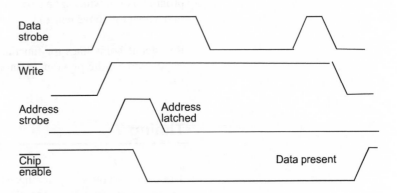

Figure 12.4 *During a read cycle, the WRITE line remains high when the RTC puts data on the bus.*

```
dirl = 255        'as outputs to bus
outl = loc        'address on bus
outc = 12     'AS low
outc = 4      'DS low
dirl = 0      'as inputs to read data
value = inl       'reading data from bus
outc = 12     'WR high
outc = 1          'disable chip

debug dec value   'displays value in register.
```

This is similar to the write program, but uses inl to load the data from the bus. The value read from the register is displayed on the computer screen.

Configuring the clock

Before the IC can be used as a clock, the values in registers A and B must be set appropriately. In Register A, UIP is a flag (update in progress), which is read-only:

UIP	DV2	DV1	DV0	RS3	RS2	RS1	RS0
0	0	1	0	0	0	0	0

DV2 to DV1 set the length of the divider chain. A table in the data sheet shows that these bits should be set to 010 for a 32.768 kHz crystal. To start with, the periodic interrupt rate and square wave output are not being used, so the RS bits are set to 0000. The write program is run with location = 10 and value = 32 (working in decimal for convenience).

Register B holds flags for functions which are of no concern at this stage, so the write program is run again with location = 11 and value = 0.

Timing

The clock is now ready for investigation, by writing values into the various locations and reading them back. To check that the clock is operating, read the seconds register (location = 0) at frequent intervals and check that it is incrementing. Similarly, the value in location 2 is found to increment once every minute. Other registers are tested very quickly by setting up the time and date as 11.59 pm on Sunday 31 December 2005, for example, and reading the data back again a few minutes later.

Square wave output

The IC has several other functions, all described in the data sheet, and we look at one of these as an example of how they can be investigated. The square wave output (pin 23) can be configured to provide output at one of a number of frequencies by setting the RS bits in Register A. For example, to obtain a 256 Hz wave, R3 is set to 1 and the other RS bits are 0. If we also need 1 in DV1 as explained above, the bits in Register A are 00101000. The values for the write program are `location` = 10 and `value` = 40. In addition, the SQWE bit (square wave enable) must be set in Register B, all other bits being 0. The SQWE bit is B3. Using the write program, make `location` = 11 and `value` = 8.

When the settings are complete, a frequency meter applied to pin 23 shows a frequency of 256 Hz.

There are several other functions in the 146818, which can all be investigated by using methods similar to those described above. It is left to the reader to work out how to implement and demonstrate these functions.

Decouple: to absorb spikes on the supply line, caused by the switching actions of the IC.

The hardware

A test bed for the IC is constructed on stripboard. It is connected to the Stamp prototyping board by two ribbon cables, one of 8 lines for the address/data bus and one of four lines for the control.

There are also two power lines, a ground (0 V) line and the regulated 5 V supply from the Stamp. A 100 μF electrolytic capacitor is connected across the supply lines where they enter the board, to decouple the supply.

A clock and pull-down circuits on pins 18 and 22 are installed as shown in the diagrams in the data sheet (see also Fig. 12.2). Pin 20 is wired to the positive supply but, as the clock output pin (pin 21) is not being used, pin 20 could equally well be connected to ground.

Activity 12.1 Investigating a support IC

Investigate the functions of an IC, following the same procedure as described above for the real time clock.

You need:

IC: This could be:

Real time clock
74HC164 SIPO
74HC165 PISO
74LS244 octal tristate buffer
74LS373 octal tristate latch
74LS574 octal D-type flip-flop
6116 or similar SRAM
RS232 output and input ICs
Programmable parallel interface (8255 PPI, 6522 VIA, M68230 PIT or the Z80 PIO)
Programmable serial interface (8250 UART, 16550 UART, M68661 DUART, M6850 ACIA, Z80 SIO)
Analogue-to-digital converter (CA3304E, ADC0804), but see Topic XX.
Digital to analogue converter
Or any other processor support IC of interest.

Data sheet.

Power supply (regulated). This may be provided by an on-board regulator or taken from an external PSU.

Microcontroller

Test bed. Depending on the system, you could build this on a breadboard, or on stripboard. Your prototyping system may have one already.

Study the data sheet and design and set up a suitable test bed. Investigate the functions of the IC by writing suitable programs.

Suggestions for investigations include:

Real time clock: Write a clock-calendar program to display date and time. Investigate the alarm function. Install a solid-state siren on the test bed and program the clock to sound it at a given time. Investigate the alarm interrupt function. Use it in a program that, say, flashes a red LED continuously but, when interrupted, flashes a green LED five times before going back to resume flashing the red LED. Set the interrupts to occur every minute. Investigate the periodic interrupt function.

74HC164, 74HC165, 74LS244, 74LS373, 74LS574 programmable interface ICs: Investigate the action of these ICs, using them as input or output ports for the processor. Connect them to input data from one or two sensors (these might be simple switches as in Fig. 2.6), and to output data to one or more actuators. The system could include address decoders to enable the port IC. Connections to the data bus must use three-state outputs. Program the system as a control system using an appropriate sensor(s) and actuator(s). Contrast the action of these ICs, suggesting to what kinds of application they are best suited. Investigate the use of handshaking and interrupt signals as detailed in the data sheet.

ADCs and DACs: Program an ADC to sample the input from a light-sensitive circuit, convert it into digital form and display it. If your system lacks a disply, program it to switch on an LED when the input voltage exceeds a given level. Program a DAC to accept a digital input. This is converted by the DAC into a varying voltage that can be used to control the speed of a small electric motor, or the brightness of a lamp.

Programming projects

This chapter concludes with a selection of programming topics. In each case, read the explanation and then write a program to demonstrate the topic.

Debouncing

Several of the programs have used the conventional routine of waiting for a keypress. An improvement is to debounce the key by using software. The routine should wait for the first contact to cause a change of input level. Then there should be a short delay, after which the input is sampled again to make sure the key is still pressed. The length of delay is important. If it is too short debouncing is not effective. If it is too long, the response is sluggish. A delay of 1 ms is reasonable as a starting point.

Lookup tables

Sometimes a value that is required may be calculated from another value by using a formula. For example, to convert a Fahrenheit temperature into a Celsius temperature we can use a formula. If it is needed several times in a program, it can be defined as a function:

```
DEF FN FtoC = (f - 32)*5/9
```

Other pairs of values may not be related in a mathematical way. For example, the groups of Morse Code are not directly related to the letters and numerals they represent. As another example, the voltage produced by a circuit using a thermistor is not directly related to the temperature.

In such cases as these we use a lookup table. If you are using assembler, this is a table in ROM, or it could be downloaded into RAM. When using a high-level language the data is stored in an array. In assembler, the table has a starting address incorporated in the program and we can access any particular value by offsetting the starting address by a given amount. For example, a table of the number of days in a month would be:

Month	RAM	Value
1	0100	31
2	0101	28
3	0102	31
4	0103	30
5	0104	31
6	0105	30
7	0106	31
8	0107	31
9	0108	30
10	0109	31
11	010A	30
12	010B	31

In assembler, given the number of the month, the stored value of the number of days can be obtained by indexed addressing. In a high-level language it can be accessed by the commands used for manipulating arrays.

It is a fairly easy project to program the month/days conversion. Slightly more difficult is to convert letters of the alphabet into the equivalent Morse Code.

Addressing modes

Microprocessors have several different ways or modes of specifying addresses, though they differ in which modes they can use. The four modes most often found are:

Direct addressing: The mnemonic is followed by the actual memory address or register where the target data is stored.

Indirect addressing: The mnemonic is followed by an address or register at which the address or register of the target data is stored.

Indexed addressing: There is an index register in the microprocessor. The mnemonic is followed by a value that is added to the contents of the index register to obtain the address at which the data is stored.

Relative addressing: The target data is stored at an address a given number of bytes further on (or further back). This is only applicable to machine code programming. A person using assembler would not be concerned with the actual addresses of the opcodes

Investigate the addressing modes available on the CPU you are using and write short routines to try them out.

Negative numbers

We usually represent a negative decimal number by writing the negation sign (-) in front of it. This does not work in a computer because the registers contain 0's or 1's, and there is no '-' symbol. There are several ways of representing negative numbers in binary without using the negation sign. The most useful of these ways is the *two's complement*.

We must first decide the number of digits in which we are working. This includes the *sign digit* on the left. We will work in four digits plus the sign digit. To form the twos complement of a number, write the positive number using 4 bits. Then write a 0 on the left to represent a positive sign. Next write the one's complement by writing 1 for every 0 and 0 for every 1. Form the two's complement by adding 1 to this number. Ignore any carry digits. This is the binary equivalent of the negative of the original number

Examples

(1) To find the 4-bit equivalent of -3.

Write +3 in binary	0011
Write the sign digit for +	00011
Find the one's complement	11100
Add 1	1
Result is two's complement	11101

This is the equivalent of -3, and the 1 on the left indicates that it is negative.

(2) Adding two negative numbers, for example, adding -2 to -3:

Two's complement of -3	11101
Two's complement of -2	11110
Add	[1]11011

Ignoring the carry digit in brackets, this is the two's complement of -5.

(3) Adding a positive number to a negative number, to get a positive result, for example adding +4 to -3.

Two's complement of -3	11101
Binary equivalent of +4	00100
Add	[1]00001

The result is +1.

(4) Adding positive to negative to get a negative result, for example, adding +2 to -6.

Two's complement of -6	11010
Binary equivalent +2	00010
Add	11100

The result is -4.

The technique also works with two positive numbers, for

example, adding +3 to +2.

Binary equivalent of +3	00011
Binary equivalent of +2	<u>00010</u>
Add	00101

The result is +5.

Most assemblers have a mnemonic for forming two's complement. Write a routine for taking two values, negating one or the other or both and calculating their sum.

Answers to questions

Chapter 1

Test your knowledge

1.1 Digital, integrates circuits, CPU, programmable.
1.2 To free the CPU for less routine tasks.
1.3 The system clock.
1.4 The CPU.
1.5 A bus.

Multiple choice questions

1 C   2 B   3 D   4 A

Chapter 2

Test your knowledge

2.1 Requires only one line.
2.2 Faster than serial transfer.
2.3 Disable all other devices with outputs connected to the data bus.
2.4 8192.
2.5 To enable a device when its address is on the address bus.
2.6 Logic high.
2.7 500 ns.
2.8 0.3 V.

Multiple choice questions

1 B   2 D   3 D   4 C   5 A   6 A   7 C

Chapter 3

Test your knowledge

1 (a) 101 1010 0100, (b) 010 1001 1111.
2 The gate is insulated from the body of the MOSFET by a layer of silicon oxide.
3 16384.
4 18.
5 A0, A2, A3, A7, A10, A12 all low; the rest high.

Multiple choice questions

1 C   2 A   3 B   4 A   5 C   6 A   7 C   8 B   9 A

Chapter 4

Test your knowledge

1 255.
2 There is an overflow and the carry flag is set to 1.
3 $Z = 0, S = 1, C = 0$.

Multiple choice questions

1 B   2 C   3 A   4 C   5 C   6 A   7 B   8 B   9 A   10 D

Chapter 5

Test your knowledge

1 1110 1010.
2 The small circle on the end of the enable line indicates active low.
3 That there is a family of devices with type numbers ranging from 74LS00 upward.
4 1111 0010.
5 F.
6 (a) 0, (b) 0, (c)1.
7 If output is 010, the outputs of the opamps must be 0000011 (from top downward), and the voltage input is 2 V (between 1.5 V and 2.5 V)
8 1.97 V.

Multiple choice questions

1 B   2 D   3 A   4 A   5 B   6 C   7 A   8 B

Chapter 7

Test your knowledge

1 So that address voltages have time to settle
2 E6 28.   3 1110 1110.

Chapter 8

Test your knowledge

1 To act like brackets enclosing titles or remarks that the CPU is to ignore.
2 This defines `counter` as an integer.
3 Defining a constant.

Chapter 9

Test your knowledge

1 The CPU would add all three and put the result in r17, leaving r16 unchanged.
2 Because there is a carry over from bits 3 to bit 4.
3 r17 would hold 48.
5 ldi r19, 15   out $17, r19.
6 ldi r20, 128   out $18,r20.
7 ldi r19, $8B   mov r17,r19.
8 r17 to r22 would all hold 20, r16 would remain unchanged.
9 right shift its digits by one place.

Chapter 10

Test your knowledge

1 The gate of the MOSFET is insulated so only a very small current is drawn from theoutput pin.
2 Z is the zero flag, it goes low when the resultof an operation is not zero.
3 4.4 ms.
4 The first number.

Index